Simon Friederich

Functional renormalization for spontaneous symmetry breaking

Simon Friederich

Functional renormalization for spontaneous symmetry breaking

An investigation of antiferromagnetism and superconductivity in the two-dimensional Hubbard model

Südwestdeutscher Verlag für Hochschulschriften

Imprint

Any brand names and product names mentioned in this book are subject to trademark, brand or patent protection and are trademarks or registered trademarks of their respective holders. The use of brand names, product names, common names, trade names, product descriptions etc. even without a particular marking in this work is in no way to be construed to mean that such names may be regarded as unrestricted in respect of trademark and brand protection legislation and could thus be used by anyone.

Publisher:
Südwestdeutscher Verlag für Hochschulschriften
is a trademark of
Dodo Books Indian Ocean Ltd., member of the OmniScriptum S.R.L Publishing group
str. A.Russo 15, of. 61, Chisinau-2068, Republic of Moldova Europe
Printed at: see last page
ISBN: 978-3-8381-2495-7

Zugl. / Approved by: Heidelberg, Universität, Dissertation, 2010

Copyright © Simon Friederich
Copyright © 2011 Dodo Books Indian Ocean Ltd., member of the OmniScriptum S.R.L Publishing group

Contents

Preface		**v**
1 Introduction		**1**
2 The Hubbard Model and the High-T_c Cuprates		**5**
2.1	High Temperature Superconductivity	5
2.2	The Hubbard Hamiltonian	7
2.3	Functional Integral Representation	8
2.4	Partial Bosonization	10
2.5	Mean Field Theory Based on Partial Bosonization	13
3 Functional Renormalization Group Formalism		**19**
3.1	Effective Action	19
3.2	Flow Equation for the Effective Average Action	21
3.3	Flowing Bosonization	23
4 Functional Renormalization for the Symmetric Regime		**25**
4.1	Truncation	26
4.2	Parameterization of Bosonic Propagators and Yukawa Couplings	29
	4.2.1 Bosonic Propagators	30
	4.2.2 Yukawa Couplings	31
4.3	Initial Conditions and Regulators	32
4.4	Flow Equations for the Running Couplings	33
	4.4.1 Flow Equations for the Yukawa Couplings	35
	4.4.2 Bosonic Propagators	49
	4.4.3 Quartic Bosonic Couplings	53
	4.4.4 Fermionic Wave Function Renormalization	55
4.5	Numerical Results	55
5 Functional Renormalization for the Spontaneously Broken Regimes		**61**
5.1	Truncation and Approximations	61
5.2	Flow of the Effective Potential	63

	5.3 Numerical results	69
	5.3.1 Order parameters	69
	5.3.2 Phase diagram	72
6	**Summary and Outlook**	**75**
A	**Notational Conventions**	**79**
B	**Pauli Matrices and Spin Projections**	**81**
C	**Box diagrams**	**83**
	C.1 Particle-particle Diagrams	83
	C.2 Particle-hole Diagrams	86
Bibliography		**91**

Preface

The present work is based on a dissertation entitled "Functional renormalization for antiferromagnetism and superconductivity in the Hubbard model" that was written at the Institute for Theoretical Physics at the University of Heidelberg between March 2008 and October 2010 under the supervision of Prof. Dr. Christof Wetterich. With respect to the dissertation, Chapters 1, 2, 3, 4, and 6 are unchanged apart from minor adjustments in some formulations and the layout of figures, whereas Chapter 5 has been substantially revised and extended. The antiferromagnetic incommensurability, which was formerly taken into account in an inadequate way, is now neglected in the regimes with spontaneous symmetry breaking, and a more thorough discussion of the temperature dependence of the order parameters for antiferromagnetism and superconductivity is included. Most of the figures shown in Chapters 4 and 5 appear also in an article, written together with Hans Christian Krahl and Christof Wetterich (for a preprint version see http://arXiv.org/abs/1012.5436), which is presently under consideration at "Physical Review B". A small number of passages in the present book is drawn from that article as well as from Refs. [35] and [36].

I am grateful to Christof Wetterich for his stimulating way of supervising my dissertation and to Jan Martin Pawlowski for answering many tricky questions that I asked him. I also would like to thank Kay-Uwe Giering, Christoph Husemann, and Andrey Katanin for many useful discussions on different topics linked to this work. A large part of the thoughts presented here in this work have emerged from a continuous dialogue with Christian Krahl. I am very grateful to him for these discussions and for the large amount of fun we had while tackling matters related to symmetry breaking and the Hubbard model. Furthermore, I would like to acknowledge the support I received from Studienstiftung des deutschen Volkes while working on my dissertation. Finally, I would like to thank my wife Andrea Harbach for listening curiously to whatever I believed I had to tell her about my work.

Bonn, March 2011

Chapter 1

Introduction

The two-dimensional Hubbard model [1, 2, 3] on a square lattice has attracted a lot of attention in the past 25 years because many researchers hope that it may throw some light on the mechanism of superconductivity in the high-T_c cuprates, which are the superconducting materials with the highest known transition temperatures from the normal to the superconducting state. In analogy to the phase diagram of the cuprates, which are antiferromagnetic at zero doping and superconducting at nonzero (either electron or hole) doping, the Hubbard model shows antiferromagnetic order at half filling and is believed to exhibit d-wave superconducting order away from half filling. Today there are many studies which predict d-wave superconductivity in a certain range of parameters aside from half filling, see e. g. [4, 5, 6, 7, 8, 9, 10, 11, 12, 13, 14, 15, 16, 17], for a systematic overview see [18].

Among the studies which were first to confirm the appearance of d-wave superconducting order in the two-dimensional Hubbard model there are some strikingly simple scaling approaches [19, 20, 21]. On a higher level of technical sophistication, the fermionic functional renormalization group approach [22, 23, 24, 25, 26, 27, 28, 29] has been of great help to analyze in detail the competition of different types of instabilities and collective order. Most studies presented so far rely on the flow of the momentum-dependent four-fermion vertex. They are performed in the so-called N-patch scheme where the Fermi surface is discretized into N patches, and the angular dependence of the four-fermion vertex is evaluated for only one momentum in each directional patch.

The approach presented in this work brings together and continues earlier attempts [30, 31, 32, 33, 34, 35, 36] to combine the advantages of the fermionic functional renormalization group with those of partial bosonization (or Hubbard-Stratonovich transformation) [37, 38]. It is based on the same version of the renormalization group idea [39, 40, 41, 42], the Wetterich flow equation for the effective average action [43], that is used in most renormalization group studies operating within a purely fermionic framework. Furthermore, it builds on the introduction of bosonic fields corresponding to different types of possible collective order of the system. The present approach is also inspired by the efficient parameterization method for the fermionic four-point vertex proposed and developed in [44]. The link between the two approaches is given by the fact that different channels of the fermionic four-point function, defined by their (almost) singular momentum structure, correspond to different types of possible orders which are described by different composite boson fields.

There are mainly two advantages of the method used in this thesis: The first is that it allows to treat the complex momentum dependence of the fermionic four-point function in an efficient, simplified way, involving only a comparatively small number of coupled flow equations. The fermionic four-point vertex, which is a scale-dependent function of three independent momenta, is decomposed in terms of bosonic propagators and Yukawa couplings, which are each functions of only one variable. A comparative disadvantage may be a better resolution of contributions from many channels in the N-patch approach. In principle, however, this disadvantage can be avoided by carefully making the choice of bosons taken into account and by choosing an appropriate parameterization for the propagators and Yukawa couplings.

The second advantage of the method used here is that it permits to follow the renormalization group flow into the phases exhibiting spontaneous symmetry breaking. (For renormalization group studies of symmetry broken phases in similar models see [45, 46, 47].) At a certain scale of the renormalization flow, the momentum-dependent fermionic four-point vertex usually diverges, and this signals the onset of local collective order. In order to extend the renormalization group treatment to the locally ordered regimes, it is necessary to describe the system in terms of composite degrees of freedom such as magnons or Cooper pairs. These are composite bosons each of which corresponds to some different type of collective order. A nonzero expectation value of the magnon field, for instance, signals the presence of some form of magnetic order, and a nonzero value of some Cooper pair field signals superconducting order. Different Cooper pair fields are distinct due to different symmetries of the order parameter they correspond to. The language of partial bosonization, where the different types of bosons are taken into account explicitly, is therefore the right tool to investigate the regimes exhibiting different forms of collective order. A particular advantage of the present approach, which combines functional renormalization and partial bosonization, is that it allows to investigate the possible coexistence of different types of order in the same range of parameters.

This thesis is structured as follows: Chapter 2 gives a brief review of superconductivity in the high-T_c cuprates and of the possible relevance of the two-dimensional Hubbard model for these materials. Afterwards, the

Hubbard model itself is reviewed and the language of partial bosonization is introduced. Based on it, a short mean field analysis of the main features of the antiferromagnetically ordered regime is given. In Chapter 3 the functional renormalization group setup is introduced on which the calculations presented in later chapters are based. Particular focus lies on the concept of the effective action Γ and on the exact flow equation for its scale-dependent relative, the effective average action or flowing action Γ_k. Chapter 4 presents the details of the renormalization group treatment for the symmetric regime. In particular, it is explained in detail how contributions to the momentum-dependent four-fermion vertex are taken into account in the partially bosonized language by means of the so-called "flowing bosonization" scheme. Flow equations and numerical results for the symmetry broken regimes are discussed in Chapter 5. In that chapter, special focus lies on the mutual influence of antiferromagnetic and d-wave superconducting order and on their possible coexistence. In Chapter 6 a brief summary of the preceding chapters is given, followed by an outlook on possible future extensions.

Chapter 2

The Hubbard Model and the High-T_c Cuprates

The present chapter starts with a brief review of high temperature superconductivity in the cuprates. The two-dimensional Hubbard Hamiltonian on a square lattice is motivated as an elementary description of the CuO_2-layers in the cuprate materials. The functional integral representation of the grand canonical partition function at finite temperature is recapitulated and the shape of the Fermi surface at small next-to-nearest neighbor hopping t' is discussed. Subsequently, I introduce the idea of partial bosonization or Hubbard-Stratonovich transformation. This designates the introduction of bosonic fields corresponding to different types of fermionic bilinears in such a manner that the four-fermion interaction is eliminated in favor of Yukawa-type interactions between the fermions and bosons. As an application of this concept, the deformation of the Fermi surface in the presence of a nonvanishing antiferromagnetic gap is derived in a mean field analysis based on the Hubbard-Stratonovich approach.

2.1 High Temperature Superconductivity

Superconductivity was discovered in 1911 by Kamerlingh Onnes in the form of a vanishing (or at least immeasurably small) electrical resistance in mercury at cryogenic temperatures. In 1933 a further fundamental characteristic of superconducting materials was discovered: According to the Meissner effect, discovered by Meissner and Ochsenfeld, the magnetic field is completely expelled from the interior of a material which undergoes a transition from the normal to the superconducting state. Both fundamental aspects of superconductivity, the vanishing resistance and the Meissner effect, were finally explained in 1957 by Bardeen, Cooper and Schrieffer (BCS) in their so-called BCS-theory of superconductivity. In this theory the transition from the normal state to a state with vanishing resistance below some nonzero critical temperature T_c is explained in terms of the formation of pairs of electrons, so-called Cooper pairs, which are the carriers of the superconducting current. The origin of Cooper pair formation in conventional superconductors is due to phonons (lattice vibrations), which explains why the critical temperatures for the transition to superconductivity are distinct for materials differing in the relative abundance of isotopes involved (isotope effect).

Estimates for the highest achievable transition temperatures from the normal to the superconducting state on the basis of the BCS-theory predicted superconductivity to be impossible for temperatures above 30 K. In April 1986, however, J. G. Bednorz and K. A. Müller discovered superconductivity in LaBaCuO, a ceramic copper oxide, with a transition temperature of approximately 30 K. Soon after, in January 1987, yttrium barium copper oxide (YBCO), another material of the same class, was discovered to have a critical temperature of 90 K, and today the highest known transition temperature of a copper oxide (at normal pressure) is as high as 135 K.

A common characteristic of all cuprate superconducting materials is their quasi two-dimensional structure. Superconductivity seems to arise from electrons moving within the weakly coupled CuO_2 layers. Other atoms and ions are confined to intermediate layers where they act as stabilizers of the three-dimensional crystal structure and as sources of additional electrons or holes: These are necessary to "dope" the copper oxide planes from the antiferromagnetically ordered, Mott insulating state which they occupy in the absence of doping into the superconducting state. The amount of (electron or hole) doping at which the highest transition temperature occurs is called "optimal doping". Similarly, one speaks of an "underdoped" ("overdoped") regime at lower (higher) than optimal doping.

An especially intriguing feature of the cuprate phase diagram is the so-called pseudogap phase, which extends above the antiferromagnetic and superconducting critical temperatures up to at least optimal doping. (For an introductory overview of the cuprate phase diagram with special emphasis on the pseudogap phase see [48].) In this regime, the arcs of the Fermi surface close to the $(\pm\pi, 0)$- and $(0, \pm\pi)$-points are destroyed so that a gap is experienced by electrons moving along the copper-oxygen bonds whereas there is no such gap for electrons moving at 45° to these bonds. The origin and properties of this phase are perhaps the least understood aspect of the cuprate phase diagram. For functional renormalization group approaches to the pseudogap phase see [49, 50, 51, 52].

Today, almost 25 years after the discovery of high temperature superconductivity, there is still no fully established consensus about the mechanism which is responsible for it. One of the most promising approaches is the spin fluctuation route to high-T_c in the cuprates, which attempts to explain cuprate superconductivity as arising from antiferromagnetic spin fluctuations in analogy to how conventional superconductivity arises from lattice vibrations [4, 5, 6, 7, 8, 9, 10]. This approach is in a sense conservative, for it takes over many essential ideas from the BCS theory of conventional superconductors, just replacing phonons by (antiferromagnetic) magnons as the "pairing glue" which is supposedly responsible for Cooper pair formation. Despite this similarity, there is, however, one crucial difference between the theory of conventional (phonon-induced) superconductivity and the spin fluctuation route to high-T_c, namely that the latter predicts the superconducting order parameter to exhibit $d_{x^2-y^2}$- rather than s-wave symmetry. Experimental evidence for the cuprates seems to confirm this prediction of the spin fluctuation approach. The relation between antiferromagnetic spin fluctuations as a possible mechanism of Cooper pair formation and the d-wave symmetry of the superconducting order parameter will be further discussed in Chapter 4.

2.2 The Hubbard Hamiltonian

In 1963 the Hubbard Hamiltonian was introduced independently by Hubbard [1], Kanamori [2] and Gutzwiller [3]. It describes fermions (e. g. electrons) on a lattice by means of a hopping term, which accounts for the motion of the fermions between different lattice sites, and an "onsite" interaction term, which accounts for the local Coulomb repulsion of fermions having opposite spin on the same lattice site. In the language of second quantization the Hamiltonian of the (one-band) Hubbard model reads

$$H = \sum_{i,j,\sigma} t_{i,j} c^\dagger_{i,\sigma} c_{j,\sigma} + U \sum_i n_{i,\uparrow} n_{i,\downarrow}, \qquad (2.1)$$

where the numbers $t_{i,j}$ account for the hopping of electrons among different lattice sites and the "Hubbard interaction" U measures the onsite repulsion. The number operator $n_{i,\sigma}$ is defined in terms of creation and annihilation operators as

$$n_{i,\sigma} = c^\dagger_{i,\sigma} c_{i,\sigma}. \qquad (2.2)$$

The physical content of the model defined through the Hamiltonian (2.1) depends on the dimensionality and topology of the underlying lattice as well as on the dimensionless ratios between the onsite repulsion U and the hopping parameters $t_{i,j}$. In this work, I shall focus on the case of the two-dimensional square lattice, which, as remarked above, is of special interest as a candidate model for the physics of the CuO_2-planes in the high-T_c suprates. The entries of the hopping matrix $t_{i,j}$ are given by three different values, namely $t_{i,j} = -t$ for neighboring lattice sites, $t_{i,j} = -t'$ for next-to-nearest neighboring lattice sites (connected by the diagonal of a square within the two-dimensional square lattice) and $t_{i,j} = 0$ for more distant lattice sites. This is sometimes referred to as the $t - t' - U$-Hubbard model.

Although the Hubbard Hamiltonian has a very simple structure, it is nevertheless able to account for a rich class of phenomena of fundamental importance, especially in solid state physics. For instance, it successfully describes the transition from a metal to a Mott insulating state and many different magnetic ordering structures such as, depending on the topology of the lattice and the choice of parameters, ferro- and antiferromagnetism. Antiferromagnetism is the dominant instability at nonzero (positive) U, vanishing next-to-nearest neighbor hopping t' and vanishing chemical potential μ, which regulates the total number of particles on the lattice.

Ever since Anderson [53] proposed the $t - t' - U$-Hubbard model on the two-dimensional square lattice as a an elementary account of the CuO$_2$-planes in the high-T_c cuprates physicists have hoped that it may throw some light on the mechanism of Cooper pairing in these materials. Anderson's proposal generated an enormous activity of trying to develop better and better approximations for the two-dimensional Hubbard model in order to improve our understanding of cuprate superconductivity. However, while the one-dimensional Hubbard model has found an analytic solution due to Lieb and Wu [54], the two-dimensional case has proved to be much less tractable. Furthermore, as experience shows, one should not expect the two-dimensional case have similar properties as the one-dimensional one. Mean field solutions, which are often a good guide in higher dimensions where each site has a sizeable number of direct neighbors, are also not very reliable in the two-dimensional case.

Among the most promising approaches are, on the one hand, numerical methods such as Monte Carlo-type approaches and, on the other hand, renormalization group accounts, for references see the Introduction (Chapter 1). The renormalization group approach used in the present work builds on the functional integral representation of the grand canonical partition function. This will be briefly reviewed in the following section.

2.3 Functional Integral Representation

The grand canonical partition function of the Hubbard model can be written as a functional integral where it reads

$$Z[\eta, \eta^\dagger] = \int_{\hat{\psi}_i^{(\dagger)}(\beta) = -\hat{\psi}_i^{(\dagger)}(0)} \mathcal{D}(\hat{\psi}, \hat{\psi}^\dagger) \exp\left(-S_F[\hat{\psi}, \hat{\psi}^\dagger] + \eta^\dagger \hat{\psi} + \eta^T \hat{\psi}^*\right) . \tag{2.3}$$

The fields $\hat{\psi}$, $\hat{\psi}^\dagger$ and the source fields η, $\eta^\dagger$ are Grassmann fields for which the notation

$$\hat{\psi}_i(\tau) = \left(\hat{\psi}_{i,\uparrow}(\tau), \hat{\psi}_{i,\downarrow}(\tau)\right)^T , \qquad \hat{\psi}_i^\dagger(\tau) = \left(\hat{\psi}_{i,\uparrow}^\dagger(\tau), \hat{\psi}_{i,\downarrow}^\dagger(\tau)\right)^T \tag{2.4}$$

is adopted and where $\hat{\psi}_i^*$ denotes the transpose of $\hat{\psi}_i^\dagger$. Furthermore, the shorthand $\eta^\dagger \hat{\psi} = \int_0^\beta \tau \sum_i \eta_i^\dagger(\tau) \hat{\psi}_i(\tau)$ is used in Eq. (2.3). (For further

2.3 Functional Integral Representation

notational conventions see Appendix A.) Using a momentum space representation, the Euclidean action S_F for the Hubbard model is given by

$$S_F = \sum_Q \hat{\psi}^\dagger(Q)[i\omega_Q + \xi_Q]\hat{\psi}(Q) \qquad (2.5)$$

$$+ \frac{U}{2} \sum_{K_1,K_2,K_3,K_4} \left[\hat{\psi}^\dagger(K_1)\hat{\psi}(K_2)\right]\left[\hat{\psi}^\dagger(K_3)\hat{\psi}(K_4)\right]\delta(K_1 - K_2 + K_3 - K_4),$$

where

$$\xi(Q) = \xi(\mathbf{q}) = -\mu - 2t(\cos q_x + \cos q_y) - 4t'\cos q_x \cos q_y \qquad (2.6)$$

is the dispersion relation for the free model on the two-dimensional square lattice.

The Fermi surface is defined for a Fermi liquid as the set of wave vectors $\mathbf{q}$ which at zero temperature separates wave vectors corresponding to occupied states from wave vectors corresponding to unoccupied states. At $U = 0$ these wave vectors are defined by the condition $\xi(\mathbf{q}) = 0$ such that states with momenta for which $\xi(\mathbf{q}) < 0$ are occupied and states with momenta for which $\xi(\mathbf{q}) > 0$ are unoccupied. However, also at finite temperature and for nonzero interactions the Fermi surface is relevant in that modes with momenta close to it give the most important contributions to scattering processes. Therefore, if contributions to scattering amplitudes have to be evaluated at selected external momenta, one should use momenta at the Fermi surface or at least close to it.

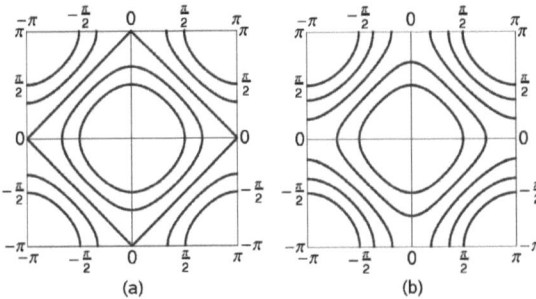

Figure 2.1: Fermi surfaces for the noninteracting Hubbard model at different values of the chemical potential $\mu/t = -2, -1, 0, 1$ and 2 (from the interior to the exterior). Fig. (a) is for vanishing next-to-nearest neighbor $t' = 0$, Fig. (b) is for $t'/t = -0.1$.

The shape and topology of the Fermi surface is crucial for the physical properties (e.g. ordering tendencies) of the system. For $t' = \mu = 0$, the

Fermi surface is a square (see Fig. 2.1 (a)) whose opposite sides can be linked through the vector $\boldsymbol{\pi} = (\pi, \pi)$. As will be discussed in Chapter 2.5, this vector is characteristic of antiferromagnetic order, which is present in the system for these values of parameters at nonzero interactions.

Insofar as the chemical potential $\mu_{1/2}$ where the lattice is half-filled (which for $t' = 0$ occurs at $\mu = 0$) can be said to correspond to zero doping, values $\mu > \mu_{1/2}$ of the chemical potential correspond to electron and values $\mu < \mu_{1/2}$ to hole doping. In the results presented in Chapters 4 and 5 focus lies on $t' < 0$ and hole doping in analogy to most cuprate superconducting materials, in particular those with the highest critical temperatures.

2.4 Partial Bosonization

Although the Hubbard model is defined as a purely fermionic model, it is useful for many purposes to introduce composite boson fields associated to certain fermion bilinears. The different types of bosons correspond to different types of possible collective order that may be characteristic of the long-range properties of the model. Formally, this idea can be spelled out by means of a scheme called *partial bosonization* or Hubbard-Stratonovich transformation, see Refs. [37, 38].

Figure 2.2: Schematic picture of bosonization of the four fermion vertex. Solid lines correspond to fermions, the dashed line to a complex (Cooper pair) boson, the wiggly line to a real boson representing a particle-hole state in the spin or charge density wave channel.

In a Hubbard-Stratonovich transformation, a purely fermionic theory is translated into a mixed theory of fermions and composite bosons, where the couplings between fermions and bosons are described by Yukawa-type vertices. The idea is graphically represented in Fig. 2.2. The first diagram on the right hand side of the arrow corresponds to a composite boson consisting of either two electrons or two holes, the second diagram to a composite (Cooper pair) boson consisting of both an electron and a hole. The two different types of bosons are used to absorb contributions to the fermionic four-point vertex in the particle-particle and particle-hole channels, respectively.

2.4 Partial Bosonization

The Hubbard interaction U can be written in terms of fermionic bilinears associated to the charge density, magnetization, and s-wave Cooper pair channels. Using a momentum space representation, the bilinears for the charge density, magnetic, s- and d-wave superconducting channels are defined by

$$\begin{aligned}
\tilde{\rho}(Q) &= \sum_P \hat{\psi}^\dagger(P)\hat{\psi}(P+Q), \\
\tilde{\mathbf{m}}(Q) &= \sum_P \hat{\psi}^\dagger(P)\boldsymbol{\sigma}\hat{\psi}(P+Q), \\
\tilde{s}(Q) &= \sum_P \hat{\psi}^T(P)\epsilon\hat{\psi}(-P+Q), \\
\tilde{s}^*(Q) &= -\sum_P \hat{\psi}^\dagger(P)\epsilon\hat{\psi}^*(-P+Q), \\
\tilde{d}(Q) &= \sum_P f_d(P-Q/2)\hat{\psi}^T(P)\epsilon\hat{\psi}(-P+Q), \\
\tilde{d}^*(Q) &= -\sum_P f_d(P-Q/2)\hat{\psi}^\dagger(P)\epsilon\hat{\psi}^*(-P+Q).
\end{aligned} \qquad (2.7)$$

Here $\boldsymbol{\sigma} = (\sigma^1, \sigma^2, \sigma^3)^T$ denotes the vector of Pauli matrices and $\epsilon = i\sigma^2$ the totally antisymmetric 2×2-tensor. The factor f_d occurring in the definition of the bilinear for the d-wave superconducting channel is the d-wave form factor

$$f_d(Q) = f_d(\mathbf{q}) = \frac{1}{2}\left(\cos(q_x) - \cos(q_y)\right). \qquad (2.8)$$

The most important qualitative features of the $d_{x^2-y^2}$-symmetry of this form factor are that it changes sign under rotations by $\pi/2$ and that its modulus is maximal at the points $\mathbf{l} = (\pi, 0)$ and $\mathbf{l}' = (0, \pi)$.

In terms of the bilinears given in Eq. (2.7) the Hubbard interaction term in Eq. (2.5) can be written in different ways, namely

$$\begin{aligned}
&\frac{U}{2} \sum_{K_1,K_2,K_3,K_4} [\hat{\psi}^\dagger(K_1)\hat{\psi}(K_2)][\hat{\psi}^\dagger(K_3)\hat{\psi}(K_4)]\delta(K_1 - K_2 + K_3 - K_4) \\
&= \frac{U}{2} \sum_Q \tilde{\rho}(Q)\tilde{\rho}(-Q) \\
&= -\frac{U}{6} \sum_Q \tilde{\mathbf{m}}(Q) \cdot \tilde{\mathbf{m}}(-Q) \\
&= \frac{U}{4} \sum_Q \tilde{s}^*(Q)s(Q).
\end{aligned} \qquad (2.9)$$

Now the basic idea of a Hubbard-Stratonovich transformation is to insert a functional integral representation of the number 1 in form of a Gaussian integral over auxiliary bosonic fields into the functional integral (2.3). These auxiliary fields are in direct correspondence to the bilinears defined in Eq. (2.7),

$$1 = \mathcal{N} \int \mathcal{D}\hat{B} \exp\left(-S_{HS}[\hat{B}]\right) = \mathcal{N} \int \mathcal{D}\hat{B} \exp\left(-S_{HS}[\hat{B} - \tilde{B}]\right), \quad (2.10)$$

where the collective boson field $\hat{B}$ refers to the collection of fields $\hat{\rho}$, $\hat{m}$, $\hat{s}$ and $\hat{s}^*$ in a vector notation,

$$\hat{B}(Q) = \left(\hat{\rho}, \hat{m}^T, \hat{s}, \hat{s}^*\right). \quad (2.11)$$

A field associated to the d-wave Cooper pair boson is not needed for the purposes of the present section, it is therefore omitted at this stage. The Hubbard-Stratonovich action S_{HS} can be chosen as

$$S_{HS}[\hat{B}] = \sum_Q \left(\frac{U_\rho}{2}\hat{\rho}(Q)\hat{\rho}(-Q) + \frac{U_m}{2}\hat{m}(Q)\cdot\hat{m}(-Q) + U_s\hat{s}^*(Q)\hat{s}(Q)\right), \quad (2.12)$$

with positive coefficients U_ρ, U_m and U_s that will be specified later. Inserting the representation of 1 defined through Eq. (2.10) in the partition function Eq. (2.3) one obtains

$$\begin{aligned} Z[j_B, \eta, \eta^\dagger] &= \mathcal{N}' \int \mathcal{D}(\hat{B}, \hat{\psi}, \hat{\psi}^\dagger) \\ &\quad \times \exp\left(-S_F[\hat{\psi}, \hat{\psi}^\dagger] - S_{HS}[\hat{B} - \tilde{B}] + j_B\hat{B} + \eta^\dagger\hat{\psi} + \eta^T\hat{\psi}^*\right), \end{aligned} \quad (2.13)$$

where $\mathcal{N}'$ is an irrelevant renormalization factor and the shorthands used are the same as in Eq. (2.3).

At this stage, an appropriate choice of the couplings U_ρ, U_m and U_s allows one to eliminate the four-fermion interaction $\sim U$, as it appears in S_F altogether and to replace it by Yukawa-type interactions between fermions and bosons. In order to achieve this, the constraint

$$U = 3U_m - U_\rho - 4U_s \quad (2.14)$$

has to be fulfilled, as follows from Eqs. (2.9) and (2.12).

In this case, the complete action $S[\hat{B}, \hat{\psi}, \hat{\psi}^\dagger] = S_F[\hat{\psi}, \hat{\psi}^\dagger] + S_{HS}[\hat{B} - \tilde{B}]$ consists of a kinetic part, including propagator terms for both bosons and fermions, and an interaction part, which accounts for the coupling between bosons and fermions.

Note that the expectation values of the fermionic bilinears and those the corresponding bosons fields are equal in this setting:

$$\langle \hat{B} \rangle = \frac{\delta}{\delta j_B} \ln Z[j_B, \eta, \eta^\dagger]\Big|_{j_b=0,\,\eta,\eta^\dagger=0} = \langle \tilde{B} \rangle. \quad (2.15)$$

The first equality follows directly from the definition of the expectation value of $\hat{B}$, the second one is obtained by first integrating out the bosons and then differentiating with respect to the source term.

In an exact treatment of the Hubbard model, all decompositions of the Hubbard interaction U that respect the constraint (2.14) would be equivalent. In practice, however, approximations have to be made, and the results obtained will in general depend on the precise choice of U_ρ, U_m and U_s, even if the choice is in accordance with Eq. (2.14). This issue, which is called the "mean field ambiguity" [55] can either be regarded as a shortcoming of the Hubbard-Stratonovich approach or as an advantage. It is a shortcoming insofar as each specific choice of the couplings U_ρ, U_m and U_s necessarily involves some bias and has an element of arbitrariness which will also appear in the results. It can be considered as advantageous, if one compares the results for different choices of the couplings and regards the robustness of the results obtained upon varying the values of the couplings as an indication of the reliability of the method one is using. The closer one gets to an exact solution of the model, the less ones results should depend on the values of the couplings U_ρ, U_m and U_s (provided the constraint (2.14) is respected).

In later chapters of the present work, however, this route will not be taken. The approach employed in the renormalization group analysis presented in Chapters 4 and 5 is not based on the elimination of the original Hubbard interaction U in a Hubbard-Stratonovich transformation, as just described. Instead, the Hubbard interaction U is kept as a four-fermion interaction on the initial ultraviolet (UV) scale $k = \Lambda$ of the renormalization flow, and contributions to the fermionic four-point vertex that arise during the flow are absorbed successively through the Yukawa couplings by means of a scale-dependent variation of the Hubbard-Stratonovich scheme. This approach, which is called *rebosonization* or *flowing bosonization*, will be described in Chapter 3.3.

In the following section, however, the Hubbard-Stratonovich method in the form just introduced is taken as the starting point for an elementary mean field approach to antiferromagnetic order in the Hubbard model. To this end, the choice $U_m/t = U/3$ and $U_\rho = U_s = 0$ is adopted in the following section.

2.5 Mean Field Theory Based on Partial Bosonization

One of the most fundamental facts about the (repulsive) Hubbard model on a square lattice is its antiferromagnetic ground state at small values of the chemical potential μ for vanishing next-to-nearest neighbor hopping t'. Later in this work (see Chapter 4.4.1) it will be explained why the tendency towards antiferromagnetism is much more pronounced than the tendencies towards charge density and s-wave superconducting order. At the present stage this is simply assumed as a known fact about the Hubbard model.

Choosing $U_m/t = U/3$ together with $U_\rho = U_s = 0$, the action S after Hubbard-Stratonovich transformation is given by

$$S[\hat{\mathbf{m}}, \hat{\psi}, \hat{\psi}^\dagger] = S_{kin}[\hat{\mathbf{m}}, \hat{\psi}, \hat{\psi}^\dagger] + S_Y[\hat{\mathbf{m}}, \hat{\psi}, \hat{\psi}^\dagger], \quad (2.16)$$

where

$$S_{kin}[\hat{\mathbf{m}}, \hat{\psi}, \hat{\psi}^\dagger] = \sum_Q \left(\hat{\psi}^\dagger(Q)[i\omega_Q + \xi_Q]\hat{\psi}(Q) + \frac{\bar{m}_m^2}{2} \hat{\mathbf{m}}(Q) \cdot \hat{\mathbf{m}}(-Q) \right) (2.17)$$

and

$$S_Y[\hat{\mathbf{m}}, \hat{\psi}, \hat{\psi}^\dagger] = -\bar{h}_m \sum_{K,Q,Q'} \delta(K - Q + Q') \hat{\mathbf{m}}(K) \cdot \hat{\psi}^\dagger(Q)\boldsymbol{\sigma}\hat{\psi}(Q') \quad (2.18)$$

with the "magnetic" mass term $\bar{m}_m^2$ and Yukawa coupling $\bar{h}_m$.

As the present study focuses on the parameter regimes of the Hubbard model where antiferromagnetism rather than ferromagnetism is the dominant magnetic instability, an "antiferromagnetic" boson field $\hat{\mathbf{a}}$ is introduced which differs from the "magnetic" boson field $\hat{\mathbf{m}}$ introduced in Chapter 2.3 through a shift in the momentum variable by the antiferromagnetic wave vector $\Pi = (0, \pi, \pi)$:

$$\hat{\mathbf{a}}(Q) = \hat{\mathbf{m}}(Q + \Pi). \quad (2.19)$$

A ferromagnetic phase does exist, but it occurs only for large values of $-t'$ requiring a different approach than the one given in this work. Together with the substitution of the **m**-boson in favor of the **a**-boson the mass term $\bar{m}_m^2$ and the Yukawa coupling $\bar{h}_m$ are replaced by m_a^2 and $\bar{h}_a$.

As a result of the Hubbard-Stratonovich transformation, the values of the antiferromagnetic mass term and Yukawa coupling can be obtained from Eqs. (2.12), (2.13), (2.17) and (2.18) as

$$\bar{m}_a^2 = \bar{h}_a = U_m, \quad (2.20)$$

which has been set to $U_m = U/3$.

In a mean field analysis of the Hubbard model after Hubbard-Stratonovich transformation the fermionic one-loop correction to the term in S_{kin} which is quadratic in the field $\hat{\mathbf{m}}$ (or, equivalently, the field $\hat{\mathbf{a}}$) depends on momentum. The corresponding coefficient in S_{kin} is the inverse antiferromagnetic propagator, which will be referred to as $\tilde{P}_a(Q)$ in what follows. The mass term $\bar{m}_a^2$ is defined as the minimal value of this inverse propagator, and the difference between $\bar{m}_a^2$ and $\tilde{P}_a(Q)$ is given by the (strictly positive) so-called kinetic term $P_a(Q)$:

$$\tilde{P}_a(Q) = \bar{m}_a^2 + P_a(Q) \quad (2.21)$$

2.5 Mean Field Theory Based on Partial Bosonization

As will be explained in detail in Chapter 4, a magnetic instability of the system, corresponding to the emergence of some type of magnetic order, occurs whenever the mass term $\bar{m}_a^2$ is reduced below zero. In accordance with Eq. (2.21), this is the case if there is some frequency-momentum Q for which the momentum-dependent 1-loop correction $\Delta P_a(Q)$ to the inverse propagator $\tilde{P}_a(Q)$ is more strongly negative than $\bar{m}_a^2$ positive, i. e. if

$$\bar{m}_a^2 + \Delta P_a(Q) \leq 0 \tag{2.22}$$

for some value of Q.

As will be discussed in detail in Chapter 4.4.2, the one-loop correction $\Delta P_a(Q)$ to the inverse antiferromagnetic propagator is given by

$$\Delta P_a(Q) = \bar{h}_a^2 \sum_P \frac{1}{P_F(Q+P+\Pi)P_F(P)} + (Q \to -Q), \tag{2.23}$$

where $P_F(Q) = i\omega_Q + \xi_Q$ denotes the fermionic propagator and $\Pi = (0, \pi, \pi)$ is the antiferromagnetic wave vector. It is introduced in Eq. (2.23) in order to ensure that the minimal value of $\tilde{P}_a(Q)$ occurs at $Q = 0$ if antiferromagnetism is the dominant instability. For the range of parameters where ferromagnetism is the dominant instability the minimum of $\tilde{P}_a(Q)$ would be situated at $Q = \Pi$.

Carrying out the Matsubara sum, which is implicit in Eq. (2.23), $\Delta P_a(Q)$ can be written as

$$\Delta P_a(\omega_Q, \mathbf{q}) = -\frac{\bar{h}_a^2}{2} \int_{[-\pi,\pi]^2} \frac{dp^2}{(2\pi)^2} \frac{\tanh(\xi_{\mathbf{p}}) - \tanh\left(\xi_{\mathbf{q}+\mathbf{p}-\pi} + \frac{i\omega_Q}{2T}\right)}{\xi_{\mathbf{p}} - \xi_{\mathbf{q}+\mathbf{p}-\pi} - i\omega_Q}$$
$$+ (Q \to -Q) \tag{2.24}$$

The remaining integral over the Brillouin zone can easily be performed numerically. It is found that the loop contribution $\Delta P_a(\omega_Q, \mathbf{q})$ is most important at zero (bosonic) Matsubara frequency $\omega_Q = 0$, so it seems reasonable to focus on the spatial momentum dependence of the $\omega_Q = 0$-contributions.

As one expects from the known facts about antiferromagnetism in the Hubbard model for $t' = 0$, at half filling and for sufficiently high temperatures also close to half filling the mean field formula Eq. (2.24) produces a pronounced minimum of $P_a(0, \mathbf{q})$ at $\mathbf{q} = 0$, see Fig. 2.3 (a). However, away from half filling the picture is different for sufficiently low temperatures, see Fig. 2.3 (b). In the center at $\mathbf{q} = 0$ there is a local maximum and there are four minima at positions

$$\mathbf{q}_{1,2} = (\pm \hat{q}, 0), \quad \mathbf{q}_{3,4} = (0, \pm \hat{q}), \tag{2.25}$$

where $\hat{q}$ is a function of T, μ, and t'. This is a manifestation of the dominance of *incommensurate* antiferromagnetic fluctuations. Experimentally,

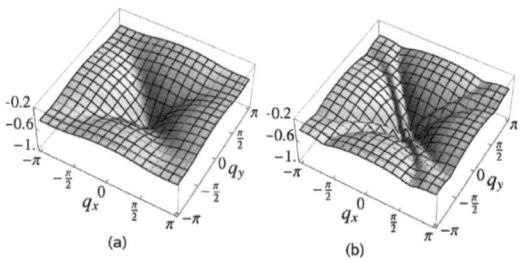

Figure 2.3: Mean field kinetic term $P_a(0,\mathbf{q})/t$ of the **a**-boson as a function of space-like momenta for $U/t = 3$ and $t' = 0$. In Figure (a) $\mu = 0$ and $T/t = 0.205$, in Figure (b) $\mu/t = -0.27$ and $T/t = 0.0435$. Both temperatures are mean field critical temperatures.

incommensurate antiferromagnetism manifests itself in the peak structure of the magnetic structure factor which is accessible via neutron-scattering. It has been observed for a variety of high-T_c-cuprates, for experimental and numerical results see [56, 57, 58, 59, 60, 61, 62].

If $P_a(0, \mathbf{q})$ has its minimal value at $\mathbf{q} = 0$, the order parameter for antiferromagnetism is given by $\langle |\mathbf{a}| \rangle \sim \delta(\mathbf{q})$, which indicates (ordinary) commensurate antiferromagnetism, exemplified by the Néel state, in which the spin direction on a given lattice site is opposite to that of its neighbors. In case, however, the minimum is located at $\mathbf{q} = \mathbf{q}_j \neq 0$, incommensurate antiferromagnetic fluctuations dominate over commensurate ones. If $\bar{m}_a^2$ drops to zero in this case, further lattice symmetries are broken. One of the pairs of minima (2.25) is selected and the symmetry of rotations by $\pi/2$ around $\mathbf{q} = 0$ in momentum space is spontaneously broken. The spins change sign between neighboring lattice sites only in one direction, the x-direction say, whereas in the orthogonal direction the periodicity corresponds to some momentum $\pi \pm \hat{q}$. Note that the system selects one of the *pairs* $\mathbf{q}_{1,2}$ or $\mathbf{q}_{3,4}$ since $\hat{a}(Q)$ is a real field. Therefore the system remains symmetric with respect to reflection about the axes.

An extensive mean field treatment of the phase with *commensurate* antiferromagnetism, including the case of a nonzero next-to-nearest neighbor hopping t', is given in [64]. Here one has to take into account that the periodicity of a system in the Néel state is changed resulting in a new "magnetic" Brillouin zone whose boundaries are given by the lines between the $(\pm \pi, 0)$ and $(0, \pm \pi)$ points. Correspondingly, the mean field dispersion relation for

2.5 Mean Field Theory Based on Partial Bosonization

a nonzero gap parameter $A = \bar{h}_a \langle |\mathbf{a}| \rangle$ has two branches

$$E_{\pm}(\mathbf{p}) = \frac{1}{2}\left(\xi(\mathbf{p}) + \xi(\mathbf{p} + \boldsymbol{\pi}) \pm \sqrt{(\xi(\mathbf{p}) - \xi(\mathbf{p} + \boldsymbol{\pi}))^2 + 4A^2}\right) \quad (2.26)$$

which, for finite t', lead to an interestingly structured effective Fermi surface enclosing hole pockets around $(\pm\pi/2, \pm\pi/2)$ and electron pockets around $(\pm\pi, 0)$ and $(0, \pm\pi)$, see the example drawn in Fig. 2.4 (a), for further details see [64].

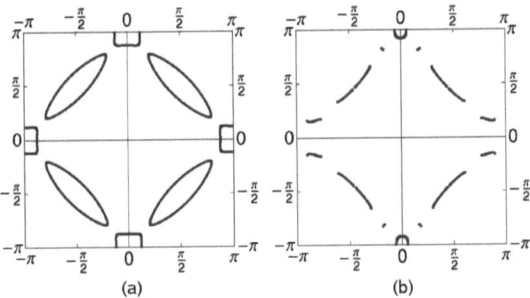

Figure 2.4: Mean field effective Fermi surfaces for $\mu/t = -0.6$, $t'/t = -0.2$ and gap parameter $A/t = 0.1$. Fig. (a) shows the commensurate case $\hat{q} = 0$ where the Fermi surface exhibits hole- and particle pockets at the magnetic Brillouin zone boundary. In Fig. (b), the remainders of the effective Fermi surface are shown for a nonzero incommensurability $\hat{q} = 0.3$ along the x-axis.

In the presence of a nonzero expectation value $\langle \mathbf{a}(\hat{\mathbf{q}}) \rangle$ with $\hat{\mathbf{q}} \neq 0$, i.e. in the presence of incommensurate order, the inverse of the fermionic mean field propagator at zero frequency has contributions from $\xi(\mathbf{q})$ but also from the gap parameter $\mathbf{A} = \bar{h}_a \langle \mathbf{a} \rangle$ and is given by

$$P_F(\mathbf{q}, \mathbf{q}') = \xi(\mathbf{q})\delta(\mathbf{q} - \mathbf{q}') \quad (2.27)$$
$$- \frac{\mathbf{A} \cdot \boldsymbol{\sigma}}{\sqrt{2}}\left(\delta(\mathbf{q} - \mathbf{q}' - \boldsymbol{\pi} + \hat{\mathbf{q}}) + \delta(\mathbf{q} - \mathbf{q}' - \boldsymbol{\pi} - \hat{\mathbf{q}})\right)$$

with $\hat{\mathbf{q}} = \mathbf{q}_{1,2}$ or $\hat{\mathbf{q}} = \mathbf{q}_{3,4}$ as defined in Eq. (2.25). The analog of the Fermi surface corresponds to the zero eigenvalues of P_F. However, the corresponding eigenmodes are no longer momentum eigenstates. Nevertheless, if the gap parameter $A = |\mathbf{A}|$ is nonzero but small, many eigenvalues of $P_F(\mathbf{q}, \mathbf{q}')$ have most of their support each at a single momentum $\mathbf{p}$. This concerns all those momenta $\mathbf{p}$ for which the condition

$$A \ll |\xi(\mathbf{p} + \boldsymbol{\pi} + \hat{\mathbf{q}})|, |\xi(\mathbf{p} + \boldsymbol{\pi} - \hat{\mathbf{q}})| \quad (2.28)$$

is fulfilled. With respect to these momenta the equation

$$\xi(\mathbf{p}) - \frac{A^2}{2}\left(\frac{1}{\xi(\mathbf{p}+\boldsymbol{\pi}+\hat{\mathbf{q}})} + \frac{1}{\xi(\mathbf{p}+\boldsymbol{\pi}-\hat{\mathbf{q}})}\right) = 0 \qquad (2.29)$$

defines an effective Fermi surface which is obtained by (approximately) diagonalizing $P_F(\mathbf{q},\mathbf{q}')$ for small A. For large enough A the effective Fermi surface vanishes completely because the number of solutions to Eq. (2.29) that satisfy the condition (2.28) diminishes rapidly. In Fig. 2.4 (b) the effective Fermi surface is shown for the incommensurate case with an order parameter $\langle \mathbf{a}(\hat{\mathbf{q}}) \rangle$ where $\hat{\mathbf{q}} = \mathbf{q}_{1,2}$, i.e. the incommensurability is along the x-axis. The symmetry of rotations by $\pi/2$ is manifestly broken.

The dichotomy between commensurate and incommensurate antiferromagnetism will also be crucial for the renormalization group treatment given in Chapters 4 and 5. For instance, the incommensurability may have an effect on whether there is a phase in which (global) antiferromagnetic and d-wave superconducting order coexist.

Chapter 3

Functional Renormalization Group Formalism

In this chapter the functional renormalization group (FRG) setup is introduced on which the calculations presented in Chapters 4 and 5 of this thesis are based. This formalism rests on an exact flow equation for the effective average action or "flowing action" Γ_k. The flowing action is a scale-dependent relative of the effective action Γ, the generating functional of one-particle irreducible (1PI) vertex functions.

In the first section of this chapter the effective action is introduced and its relation to thermodynamical quantities is clarified. An introduction to the exact flow equation for the flowing action is given in the second section. The third section presents the "flowing bosonization" scheme, a continuous, scale-dependent variation of the Hubbard-Stratonovich transformation.

3.1 Effective Action

Using the functional integral formulation of QFT, the grand-canonical partition function

$$Z = \text{Tr} \exp\left(-\beta(\text{H} - \mu \text{N})\right) \tag{3.1}$$

can be written as

$$Z[J] = \int \mathcal{D}\hat{\chi} \exp\left(-S[\hat{\chi}] + \int_X J(X) \cdot \hat{\chi}(X)\right), \tag{3.2}$$

where the chemical potential μ has been included in the source term for which the shorthand $J \cdot \hat{\chi} = \int_X J(X) \cdot \hat{\chi}(X)$ has been used.

In applications used in this work, the field $\hat{\chi}$ will be a multi-component field with both bosonic and fermionic entries. Its precise form will be specified in the following chapter when the details of the approximation ("truncation") for the effective action are discussed. The multi-component field J is defined as the current-field associated to $\hat{\chi}$.

In terms of $Z[J]$ one defines

$$W[J] = \ln Z[J], \tag{3.3}$$

the generating functional of the *connected* Green functions without external lines. The classical fields χ associated to the quantum fields $\hat{\chi}$ are defined as the expectation values of the quantum fields:

$$\chi := \langle \hat{\chi} \rangle = \frac{1}{Z[J]} \frac{\delta Z[J]}{\delta J} = \frac{\delta W[J]}{\delta J}. \tag{3.4}$$

In terms of these quantities the effective action Γ is defined through a Legendre transform from $W[J]$ as

$$\Gamma[\chi] = \sup_J \left(J \cdot \chi - W[J] \right). \tag{3.5}$$

The definition of $\Gamma[\chi]$ as a Legendre transform guarantees its convexity. It generates all (amputated) Green functions that cannot be taken apart by cutting just one internal line. In other words, $\Gamma[\chi]$ is the generating functional of the one-particle irreducible (1PI) vertex functions.

In an implicit form the effective action is given through

$$\exp(-\Gamma[\chi]) = \int \mathcal{D}\hat{\chi} \exp\left(-S[\chi + \hat{\chi}] + \frac{\delta \Gamma}{\delta \chi} \cdot \hat{\chi} \right). \tag{3.6}$$

Performing a saddle-point approximation of the classical action $S[\chi + \hat{\chi}]$ with respect to the fluctuation field $\hat{\chi}$ around the background field χ one obtains the one-loop equation

$$\Gamma[\chi] = S[\chi] + \frac{1}{2} \text{STr} \ln S^{(2)}[\chi] + \dots, \tag{3.7}$$

where "$S^{(2)}[\chi]$" denotes the second functional derivative of S with respect to the field χ and "STr" denotes the "supertrace", where the trace is taken over all indices (e.g. field, momentum and spin indices) with an additional minus-sign for fermionic entries. The flow equation for the effective average action that will be presented in the following section can be regarded as a renormalization enhanced non-perturbative version of Eq. (3.7).

If both the current field $J = j/T$ (where T is the temperature) and the classical field χ are homogeneous in space and time, all thermodynamical quantities can easily be expressed in terms of Γ. First, one defines the effective potential $U = T\Gamma/V$, and from this one obtains, for instance, the equations

$$\epsilon = U - T\frac{\partial U}{\partial T} - \mu \frac{\partial U}{\partial \mu}, \tag{3.8}$$

$$n = -\frac{\partial U}{\partial \mu}, \qquad p = -U$$

3.2 Flow Equation for the Effective Average Action

for the energy density ϵ, particle density n and pressure p, see Chapter 2.1 in [66].

Having precise knowledge of $\Gamma[\chi]$ in a concrete case means having solved the quantum many-body or field-theoretical problem at issue. In most cases, however, an exact calculation of $\Gamma[\chi]$ is impossible. One has to resort to an approximation scheme such as, for instance, the functional renormalization group approach described in the next section.

3.2 Flow Equation for the Effective Average Action

Renormalization group methods are a powerful non-perturbative tool in quantum field theory and many-body statistical physics. The Wilsonian renormalization group idea [39, 40, 41], is to integrate out quantum fluctuations step by step instead of all at once. In this work, the renormalization group idea is used within the framework of the so-called exact or functional renormalization group, which is based on an exact flow equation for the effective average action or flowing action Γ_k, a scale-dependent relative of the effective action Γ. Intuitively, Γ_k is obtained from Γ by starting from the microscopic "classical" action S and integrating out all quantum fluctuations down to some finite momentum scale k.

The microphysical properties of the system, the Hamiltonian or Lagrangian which characterize the model one is interested in, are taken into account by the condition

$$\Gamma_{k=\Lambda} = S, \tag{3.9}$$

where Λ denotes some very large UV scale and S is the microscopic action. In the infrared (IR) limit $k \to 0$ the flowing action Γ_k equals the full effective action

$$\Gamma_{k=0} = \Gamma, \tag{3.10}$$

which, as remarked above, is the generating functional of the 1PI vertex functions.

In order to obtain the flow equation for the flowing action, one first introduces a scale dependence in form of an IR regulator term ΔS_k in the classical action S:

$$S[\hat{\chi}] \to S_k[\hat{\chi}] = S[\hat{\chi}] + \Delta S_k[\hat{\chi}]. \tag{3.11}$$

The regulator term is quadratic in the fields. In terms of a regulator function R_k, which is a matrix in field space, it can be written as

$$\Delta S_k[\hat{\chi}] = \frac{1}{2} \int_Q \hat{\chi}^T(-Q) R_k(Q) \hat{\chi}(Q). \tag{3.12}$$

In order to function as an infrared cutoff, the regulator function R_k has to obey the conditions

$$\lim_{Q^2/k^2 \to 0} R_k(Q) > 0,$$
$$\lim_{k^2/Q^2 \to 0} R_k(Q) = 0, \qquad (3.13)$$
$$R_k(Q) \to \infty \text{ for } k^2 \to \Lambda.$$

These conditions ensure that Γ_k has the appropriate limiting properties as specified in Eqs. (3.9) and (3.10).

The scale dependence of the classical action introduced in Eq. (3.11) induces a scale dependence in the generating functional $W_k[J]$,

$$W_k[J] = \ln Z_k[J] = \ln \int \mathcal{D}\hat{\chi} \exp\left(-S_k[\hat{\chi}] + J \cdot \hat{\chi}\right). \qquad (3.14)$$

In order for the flow equation for the flowing action Γ_k to take a simple form, it is defined as a modified Legendre transform of W_k, namely

$$\Gamma_k[\chi] = \sup_J \left(J \cdot \chi - W_k\right) - \Delta S_k[\chi]. \qquad (3.15)$$

The exact flow equation for Γ_k governs its scale dependence in terms of the regulator R_k together with the inverse propagator, which is given by

$$\Gamma_k^{(2)}(Q, Q') = \frac{\delta^2 \Gamma_k[\chi]}{\delta\chi(-Q)\delta\chi(Q')}. \qquad (3.16)$$

The necessary ingredients for the formulation of the Wetterich flow equation for the flowing action have now all been defined. It is given by

$$\partial_k \Gamma_k = \frac{1}{2} \text{STr} \left(\Gamma_k^{(2)} + R_k\right)^{-1} \partial_k R_k = \frac{1}{2} \text{STr}\, \tilde{\partial}_k \left(\ln(\Gamma_k^{(2)} + R_k)\right). \qquad (3.17)$$

Here STr denotes a "supertrace" which sums over all quantum numbers including the different field indices with an additional minus-sign for fermionic entries. The operator $\tilde{\partial}_k = (\partial_k R_k)\frac{\partial}{\partial R_k}$ can be described as the scale derivative acting only onto the infrared regulator R_k. Eq. (3.17) is derived by first performing the derivative of W_k with respect to k and then using the properties of W_k and Γ_k as (modified) Legendre transforms of each other. For the original version of the derivation see [43], for more recent presentations see e. g. [65, 66, 67].

In the present work, a generalization of Eq. (3.17) will be used which applies also to situations where the fields χ themselves depend on the renormalization scale k. For this purpose one has to take the generalized form

$$\partial_k \Gamma_k[\chi_k]\big|_k = \partial_k \Gamma_k[\chi_k] - \int \left(\partial_k \chi_{k,i} \frac{\delta \Gamma_k[\chi_k]}{\delta \chi_{k,i}}\right), \qquad (3.18)$$

of Eq. (3.17), for a derivation and motivation see [68]. Eq. (3.18) includes a summation over the doubly occurring index i. For a more recent generalization of Eq. (3.17) see [69].

3.3 Flowing Bosonization

Flowing bosonization [67, 68, 69] can be described as a scale-dependent Hubbard-Stratonovich transformation which is carried out continuously on all scales of the renormalization flow. In the context of the present investigation it is employed to describe all parts of the fermionic four-point vertex which have a nontrivial momentum dependence in terms of effective interactions between fermions and different types of bosons.

To illustrate how flowing bosonization works, consider, as an example, a truncated theory in which the effective average action includes, besides the fermions, only one type of boson, namely the antiferromagnetic boson **a** that was introduced in Eq. (2.19). The result that will be obtained carries over to other types of bosons and is in no way specific to the **a**-boson.

Consider an ansatz for the effective average action given by

$$\Gamma_{a,k} + \Gamma_{Fa,k} + \Gamma_{F,k}^{a} = \tag{3.19}$$

$$\frac{1}{2}\sum_{Q} \mathbf{a}^{T}(-Q)\left(P_a(Q) + m_a^2\right)\mathbf{a}(Q)$$

$$-\sum_{K,Q,Q'} \bar{h}_a(K)\,\mathbf{a}(K) \cdot [\psi^\dagger(Q)\boldsymbol{\sigma}\psi(Q')]\,\delta(K-Q+Q'-\Pi)$$

$$-\frac{1}{2}\sum_{K_1,K_2,K_3,K_4} \lambda_F^a(K_1-K_2)\delta\left(K_1-K_2+K_3-K_4\right)$$

$$\times \left[\psi^\dagger(K_1)\boldsymbol{\sigma}\psi(K_2)\right] \cdot \left[\psi^\dagger(K_3)\boldsymbol{\sigma}\psi(K_4)\right].$$

Now a scale dependence of the field $\mathbf{a}(Q)$ is introduced, writing it as $\mathbf{a}_k(Q)$. The change in $\mathbf{a}_k(Q)$ between two scales k and $k-\Delta k$ that are infinitesimally close to each other can be represented as

$$\mathbf{a}_k(Q) - \mathbf{a}_{k-\Delta k}(Q) = \Delta\alpha_k(Q)\tilde{\mathbf{a}}_k(Q), \tag{3.20}$$

where the field $\tilde{\mathbf{a}}_k(Q)$ is given by the fermion bilinear

$$\tilde{\mathbf{a}}_k(Q) = \sum_P [\psi^\dagger(P)\boldsymbol{\sigma}\psi(P+Q+\Pi)], \tag{3.21}$$

and $\alpha_k(Q)$ is a function that can be chosen in such a way that λ_F^a cancels to zero at all scales.

To achieve this, the generalized flow equation Eq. (3.18) in terms of scale-dependent fields is taken, which yields in this case

$$\partial_k \Gamma_k = \partial_k \Gamma_k \big|_{\mathbf{a}_k} + \sum_Q \Big(\partial_k \alpha_k(Q)\tilde{P}_a(Q)\mathbf{a}_k(-Q) \cdot \tilde{\mathbf{a}}_k(Q)$$

$$- \partial_k \alpha_k(Q)\bar{h}_a(Q)\tilde{\mathbf{a}}_k(-Q) \cdot \tilde{\mathbf{a}}_k(Q) \Big). \tag{3.22}$$

One can read off the modified equations for λ_F^a and $\bar{h}_a$ and set the scale-dependence of λ_F^a to zero,

$$\partial_k \bar{h}_a(Q) = \partial_k \bar{h}_a\big|_{\mathbf{a}_k}(Q) - \tilde{P}_a(Q)\partial_k \alpha_k(Q), \qquad (3.23)$$
$$\partial_k \lambda_F^a(Q) = \partial_k \lambda_F^a\big|_{\mathbf{a}_k}(Q) + 2\bar{h}_a(Q)\partial_k \alpha_k(Q) \equiv 0.$$

This allows one to eliminate the hitherto undetermined function $\alpha_k(Q)$ and to obtain the flow equation for the Yukawa coupling including contributions from flowing bosonization,

$$\partial_k \bar{h}_a(Q) = \partial_k \bar{h}_a\big|_{\mathbf{a}_k}(Q) + \frac{\tilde{P}_{a,k}(Q)}{2\bar{h}_a(Q)}\partial_k \lambda_F^a\big|_{\mathbf{a}_k}(Q). \qquad (3.24)$$

Analogous results can be obtained for the four-fermion couplings and Yukawa couplings in other than the magnetic channel such as, for instance, the charge density and superconducting channels, see Eq. (4.38). The result Eq. (3.24) will be of crucial importance in the derivation of the flow equations for the Yukawa couplings presented in the next section.

Chapter 4

Functional Renormalization for the Symmetric Regime

Although the Wetterich flow equation Eq. (3.17) for the flowing action Γ_k is an exact equation, it is in general not possible to use it for exact solutions to difficult problems of physical interest. Since Γ_k is a functional of the (many-component) field χ Eq. (3.17) implicitly contains differential equations for an infinity of running couplings so that it can mostly be solved only in some approximation. For practical purposes, an ansatz for the flowing action in terms of a system of running parameters has to be specified and the resulting (infinite) hierarchy of coupled flow equation has to be *truncated* at some point. In other words, a finite number of well-chosen couplings has to be selected so that the resulting system of coupled differential equation can be solved either analytically or numerically. The truncation used in the present work for the regime where no symmetry of the Hubbard Hamiltonian is spontaneously broken is specified in the following section of the present chapter.

Most functional renormalization group investigations of the Hubbard model use the purely fermionic language, in which the Hubbard model itself is formulated [22, 23, 24, 25, 26, 27, 28, 29]. In this work, in contrast, the partly bosonized language introduced in Chapter 2.3 is employed, which greatly facilitates the treatment of phases exhibiting spontaneous symmetry breaking. The Hubbard-Stratonovich transformation, however, is not carried out for the original Hubbard action (2.5) but rather continuously on all scales of the renormalization flow by using the method of flowing bosonization described in Chapter 3.3.

Since the complicated spin and momentum dependence of the four-fermion coupling which arises during the renormalization flow is captured in the present approach in terms of bosonic propagators and Yukawa couplings, the parameterizations for these have to be chosen with great care. How this is done is described in the second section of this chapter. Having chosen a truncation and parameterization for the running couplings, regulator terms for both fermions and bosons and initial conditions for the running couplings have to be specified. This is done in the third section of this chapter. The fourth section derives and discusses the flow equations for the running couplings obtained in this setting, the fifth summarizes the most important numerical results obtained for the symmetric regime.

4.1 Truncation

To begin with, an ansatz for the flowing action has to be specified. As argued before, it should include terms for the electrons (fermions), for bosons in the magnetic, charge density, s- and d-wave superconducting channels, and for interactions between fermions and bosons. Such an ansatz may have the form

$$\Gamma_k[\chi] = \Gamma_{F,k} + \Gamma_{Fa,k} + \Gamma_{F\rho,k} + \Gamma_{Fs,k} + \Gamma_{Fd,k} \quad (4.1)$$
$$+ \Gamma_{a,k} + \Gamma_{\rho,k} + \Gamma_{s,k} + \Gamma_{d,k} + \sum_X U_k(\mathbf{a}, \rho, s, d),$$

where the meaning of the terms on the right hand side will be specified in what follows. The collective field $\chi = (\mathbf{a}, \rho, s, s^*, d, d^*, \psi, \psi^*)$ includes both fermion fields ψ, ψ^* and boson fields $\mathbf{a}, \rho, s, s^*, d, d^*$.

The purely fermionic part Γ_F (the dependence on the scale k is always implicit in what follows) of the flowing action consists of a two-fermion kinetic term $\Gamma_{F\text{kin}}$, a momentum-independent four-fermion term Γ_F^U, and the momentum-dependent four-fermion terms $\Gamma_F^m, \Gamma_F^\rho, \Gamma_F^s, \Gamma_F^d$:

$$\Gamma_F = \Gamma_{F\text{kin}} + \Gamma_F^U + \Gamma_F^m + \Gamma_F^\rho + \Gamma_F^s + \Gamma_F^d. \quad (4.2)$$

The fermionic kinetic term is essentially left unchanged with respect to the original (microscopic) action of the Hubbard model (2.5), apart from the fact that a fermionic wave function renormalization is included, which depends on the Matsubara frequency,

$$\Gamma_{F\text{kin}} = \sum_Q \psi^\dagger(Q) P_F(Q) \psi(Q), \quad (4.3)$$

where as an ansatz for the inverse fermionic propagator

$$P_F(Q) = Z_F(\omega_Q)(i\omega_Q + \xi(\mathbf{q})) \quad (4.4)$$

is made with a frequency-dependent wave function renormalization factor $Z_F(\omega_Q)$.

The flow of $Z_F(\omega_Q)$ is neglected for all frequencies except for the two lowest Matsubara modes $\omega_Q = \pm \pi T$. The computation of the scale-dependent

4.1 Truncation

quantity $Z_F(\pm\pi T) \equiv Z_F$ is described in Chapter 4.4. Self-energy corrections to the dependence of $P_F(Q)$ on spatial momentum are omitted. According to [70] their influence is small compared to that of corrections to the frequency dependence of $P_F(Q)$ for the lowest modes.

The momentum-independent part of the four-fermion coupling, which at $k = \Lambda$ is identical to the Hubbard interaction U, remains unmodified during the flow. The corresponding part of the effective action has the same form as the interaction part of the original action S of the Hubbard model, so it reads

$$\begin{aligned}\Gamma_F^U &= \frac{1}{2} \sum_{K_1,K_2,K_3,K_4} U\,\delta\,(K_1 - K_2 + K_3 - K_4) \\ &\quad \times \left[\psi^\dagger(K_1)\psi(K_2)\right]\left[\psi^\dagger(K_3)\psi(K_4)\right].\end{aligned} \quad (4.5)$$

Most information about instabilities and ordering tendencies of the system is contained in the complicated momentum and spin dependence of the fermionic four-point function $\lambda_F(K_1, K_2, K_3, K_4)$, which, due to energy-momentum conservation, is a function of three independent momenta (e. g. $K_4 = K_1 - K_2 + K_3$). In the truncation used here this vertex is decomposed into a sum of four functions $\lambda_F^a(Q)$, $\lambda_F^\rho(Q)$, $\lambda_F^s(Q)$ and $\lambda_F^d(Q)$, each depending on only one particular combination of the K_i. The chosen decomposition of the fermionic four-point function is inspired by the singular frequency and momentum structure of the leading contributions during the renormalization flow. In the ansatz for the effective average action used here these functions enter as

$$\begin{aligned}\Gamma_F^m = -\frac{1}{2} \sum_{K_1,K_2,K_3,K_4} &\lambda_F^a(K_1 - K_2)\,\delta\,(K_1 - K_2 + K_3 - K_4) \\ &\times \left[\psi^\dagger(K_1)\boldsymbol{\sigma}\psi(K_2)\right] \cdot \left[\psi^\dagger(K_3)\boldsymbol{\sigma}\psi(K_4)\right],\end{aligned} \quad (4.6)$$

$$\begin{aligned}\Gamma_F^\rho = -\frac{1}{2} \sum_{K_1,K_2,K_3,K_4} &\lambda_F^\rho(K_1 - K_2)\,\delta\,(K_1 - K_2 + K_3 - K_4) \\ &\times \left[\psi^\dagger(K_1)\psi(K_2)\right]\left[\psi^\dagger(K_3)\psi(K_4)\right]\end{aligned} \quad (4.7)$$

for the real bosons, and, for the superconducting bosons, as

$$\begin{aligned}\Gamma_F^s = \sum_{K_1,K_2,K_3,K_4} &\lambda_F^s(K_1 + K_3)\,\delta\,(K_1 - K_2 + K_3 - K_4) \\ &\times \left[\psi^\dagger(K_1)\epsilon\psi^*(K_3)\right]\left[\psi^T(K_2)\epsilon\psi(K_4)\right],\end{aligned} \quad (4.8)$$

$$\begin{aligned}\Gamma_F^d = \sum_{K_1,K_2,K_3,K_4} &\lambda_F^d(K_1 + K_3)\,\delta\,(K_1 - K_2 + K_3 - K_4) \\ &\times f_d((K_1 - K_3)/2)\,f_d((K_2 - K_4)/2) \\ &\times \left[\psi^\dagger(K_1)\epsilon\psi^*(K_3)\right]\left[\psi^T(K_2)\epsilon\psi(K_4)\right].\end{aligned} \quad (4.9)$$

In the partially bosonized approach used here all information contained in the momentum dependence of the couplings λ_F^a, λ_F^ρ, λ_F^s and λ_F^d is expressed in terms of Yukawa couplings and bosonic propagators. Practically, this is achieved by the technique of flowing bosonization, which was adapted to the purposes of the present investigation in [31, 34, 36]. As described in Chapter 3.3, the basic idea is to introduce scale-dependent bosonic fields in order to reexpress all information contained in the fermionic four-point vertex in terms of Yukawa-type interactions between fermions and bosons. In this way all couplings λ_F^a, λ_F^ρ, λ_F^s and λ_F^d are kept vanishing during the flow, and their k-dependence is absorbed by that of the flowing Yukawa couplings describing the interactions between fermions and bosons. The parts of the truncation in which these couplings occur are given by

$$\Gamma_{Fa} = -\sum_{K,Q,Q'} \bar{h}_a(K)\, \mathbf{a}(K) \cdot [\psi^\dagger(Q)\boldsymbol{\sigma}\psi(Q')]\, \delta(K - Q + Q' + \Pi),$$

$$\Gamma_{F\rho} = -\sum_{K,Q,Q'} \bar{h}_\rho(K)\, \rho(K)\, [\psi^\dagger(Q)\psi(Q')]\, \delta(K - Q + Q'),$$

$$\Gamma_{Fs} = -\sum_{K,Q,Q'} \bar{h}_s(K)\, \big(s^*(K)\, [\psi^T(Q)\epsilon\psi(Q')] \qquad (4.10)$$

$$-s(K)\, [\psi^\dagger(Q)\epsilon\psi^*(Q')]\big)\, \delta(K - Q - Q'),$$

$$\Gamma_{Fd} = -\sum_{K,Q,Q'} \bar{h}_d(K) f_d\left((Q-Q')/2\right) \big(d^*(K)\, [\psi^T(Q)\epsilon\psi(Q')]$$

$$-d(K)\, [\psi^\dagger(Q)\epsilon\psi^*(Q')]\big)\, \delta(K - Q - Q').$$

The purely bosonic part of the truncation for the effective average action consists of the bosonic kinetic terms together with the bosonic effective potential. As discussed for the inverse antiferromagnetic propagator in Chapter 2.5, the inverse propagator of some boson $i = a, \rho, s, d$ is given by $\tilde{P}_i(Q) \equiv P_i(Q) + \bar{m}_i^2$, where $\bar{m}_i^2$ is its minimal value and $P_i(Q)$ the (strictly positive) kinetic term. The contributions to the effective average action where the bosonic kinetic terms appear are

$$\Gamma_a = \frac{1}{2}\sum_Q \mathbf{a}^T(-Q) P_a(Q) \mathbf{a}(Q), \qquad (4.11)$$

$$\Gamma_\rho = \frac{1}{2}\sum_Q \rho(-Q) P_\rho(Q) \rho(Q), \qquad (4.12)$$

$$\Gamma_s = \sum_Q s^*(Q) P_s(Q) s(Q), \qquad (4.13)$$

$$\Gamma_d = \sum_Q d^*(Q) P_d(Q) d(Q). \qquad (4.14)$$

The parameterization used for the frequency- and momentum-dependence

4.2 Parameterization of Bosonic Propagators and Yukawa Couplings

of the inverse bosonic propagators $\tilde{P}_i(Q)$ and Yukawa couplings $\hat{h}_i(Q)$ will be specified in the following section.

One can reconstruct the momentum-dependent four-fermion interactions Γ_F^i by solving the field equation for the bosons i as a functional of fermionic variables (as derived by variation of Γ_k with respect to the field for the i-boson) and reinserting this functional into Γ_k. So, the complicated spin and momentum dependence of the fermionic four-point funcation which emerges during the renormalization flow is completely expressed by the bosonic propagators and Yukawa-couplings connecting the fermions to the different bosons.

The truncation also includes a local effective potential $U(\mathbf{a}, \rho, s, d)$ (not to be confused with the Hubbard interaction U, for which, due to convention, the same letter 'U' is also used). Here an expansion in powers of fields $\mathbf{a}$, ρ, s, and d is made up to second order in ρ and s and up to fourth order in $\mathbf{a}$ and d. Spontaneous symmetry breaking in the antiferromagnetic or d-wave superconducting channel can be described in terms of a minimum of U the position of which does not coincide with the origin in $\mathbf{a}$-d-space. A sufficient condition for spontaneous symmetry breaking is that the terms in the effective potential which are quadratic in $\mathbf{a}$ and d turn negative. In the symmetric regime SYM, where one has positive mass terms $\bar{m}_a^2$ and $\bar{m}_d^2$, the effective potential can be expanded around the origin in $\mathbf{a}$-ρ-s-d-space:

$$\begin{aligned}
\sum_X U(\mathbf{a}, \rho, s, d) &= \sum_Q \{ \tfrac{1}{2} \left(\bar{m}_a^2 \, \mathbf{a}^T(-Q) \mathbf{a}(Q) + \bar{m}_\rho^2 \, \rho(-Q)\rho(Q) \right) \\
&\quad + \bar{m}_s^2 \, s^*(Q)s(Q) + \bar{m}_d^2 \, d^*(Q)d(Q) \} \\
&\quad + \tfrac{1}{2} \sum_{Q_1, Q_2, Q_3, Q_4} \delta(Q_1 + Q_2 + Q_3 + Q_4) \\
&\quad \times \left(\bar{\lambda}_a \, \alpha(Q_1, Q_2)\alpha(Q_3, Q_4) + \bar{\lambda}_d \, \delta(Q_1, Q_2)\delta(Q_3, Q_4) \right. \\
&\quad \left. + 2\bar{\lambda}_{ad} \, \alpha(Q_1, Q_2)\delta(Q_3, Q_4) \right) \,.
\end{aligned} \quad (4.15)$$

Here the transformation-invariant quantities $\alpha(Q_1, Q_2) = \tfrac{1}{2}\mathbf{a}(Q_1) \cdot \mathbf{a}(Q_2)$ and $\delta(Q_1, Q_2) = d^*(Q_1)d(Q_2)$ have been used. The symbol $\delta(Q_1, Q_2)$ is to be distinguished from the Dirac delta-function by the number of arguments.

4.2 Parameterization of Bosonic Propagators and Yukawa Couplings

In the language of partial bosonization the momentum-dependent parts λ_F^i of the fermionic four-point vertex are set to zero at the expense of additional contributions to the bosonic propagators and Yukawa couplings. In principle, by slightly varying the flowing bosonization scheme presented in Chapter 3.3, the momentum dependence of the λ_F^i could be absorbed by a momentum dependence in either the $\bar{h}_i$ or the $\tilde{P}_i$, leaving the other momentum-independent. For instance, it may seem natural to keep the Yukawa couplings momentum-independent while endowing the inverse propagators $\tilde{P}_i$ with a momentum dependence. Computationally, however, it is more convenient to treat both the $\tilde{P}_i$ and the $\bar{h}_i$ as momentum-dependent functions, in addition to their dependence on the scale k.

In a numerically more exact partially bosonized treatment of the fermionic four-point vertex one would discretize the momentum dependence and attempt a numerical solution of the partial differential equations for $\tilde{P}_i(Q)$ and $\bar{h}_i(Q)$ (see the numerical treatment of the "bosonic propagators" in [44]). However, since the present approach aims at a computationally economical approach that focuses on physical understanding rather than quantitative accuracy, a parameterization of the Yukawa couplings and inverse propagators in terms of only a few running parameters is chosen. This is described in what follows.

4.2.1 Bosonic Propagators

For the antiferromagnetic kinetic term $P_a(Q)$ an ansatz is made in which the dependences on frequency and spatial momentum decouple,

$$P_a(Q) = Z_a \omega_Q^2 + A_a F(\mathbf{q}) . \tag{4.16}$$

The quadratic dependence on frequency is motivated by mean field results for small $|\omega_Q|$. For larger values of $|\omega_Q|$, it mimics the decaying frequency-dependence of the Yukawa couplings, which is not taken into account explicitly.

For the function $F(\mathbf{q})$ in Eq. (4.16) the parameterization

$$F_c(\mathbf{q}) = \frac{D_a^2 \cdot [\mathbf{q}]^2}{D_a^2 + [\mathbf{q}]^2} \tag{4.17}$$

is chosen, if commensurate antiferromagnetic fluctuations dominate, see the Appendix to [34]. Here $[\mathbf{q}]^2$ is defined as $[\mathbf{q}]^2 = q_x^2 + q_y^2$ for $q_{x,y} \in [-\pi, \pi]$ and continued periodically otherwise. If incommensurate antiferromagnetic fluctuations dominate, the alternative ansatz

$$F_i(\mathbf{q}, \hat{q}) = \frac{D_a^2 \tilde{F}(\mathbf{q}, \hat{q})}{D_a^2 + \tilde{F}(\mathbf{q}, \hat{q})} \tag{4.18}$$

is used where the function $\tilde{F}$ has a quartic momentum dependence and explicitly includes the incommensurability $\hat{q}$:

$$\tilde{F}(\mathbf{q}, \hat{q}) = \frac{1}{4\hat{q}^2}\left((\hat{q}^2 - [\mathbf{q}]^2)^2 + 4[q_x]^2[q_y]^2\right) . \tag{4.19}$$

4.2 Parameterization of Bosonic Propagators and Yukawa Couplings 31

The first term in $\tilde{F}$ vanishes for $[\mathbf{q}]^2 = \hat{q}^2$ and suppresses the propagator for $[\mathbf{q}]^2 \neq \hat{q}^2$. The second term favors the minima at $\mathbf{q}_{1,2}$ and $\mathbf{q}_{3,4}$ as compared to a situation where rotation-symmetry in the $q_x - q_y$-plane is preserved.

The shape coefficient D_a used in Eqs. (4.17), (4.18) is computed as

$$D_a^2 = \frac{1}{A_a}\bigl(P_a(0,\pi,\pi) - P_a(0,\hat{q},0)\bigr). \tag{4.20}$$

The Z_a- and A_a-factors are computed from the differences of inverse propagators at different frequencies and momenta around the minimal value of P_a,

$$\begin{aligned}
Z_a &= \frac{1}{(2\pi T)^2}\left(P_a(2\pi T, \hat{q}, 0) - P_a(0, \hat{q}, 0)\right), \\
A_a &= \frac{1}{\bar{q}^2}\left(P_a(0, \hat{q}+\bar{q}, 0) - P_a(0, \hat{q}, 0)\right),
\end{aligned} \tag{4.21}$$

where $\bar{q}$ is a parameter which is fixed in such a way that results are practically independent of it. For the numerical results displayed later, it is set to $\bar{q} = 0.15\pi$.

The propagator of the ρ-boson is treated in the same way as that of the **a**-boson apart from the fact that a shift by the vector $\boldsymbol{\pi}$ is included in all momentum dependences.

For the s-and d-bosons, the treatment is just as for the a- and ρ-bosons in the commensurate case. Since the minima of the inverse propagator do not occur at the wave vector $\boldsymbol{\pi}$ nor in the vicinity of it, no shift by this vector is necessary,

$$P_{s/d,k}(Q) = Z_{s/d}\omega^2 + A_{s/d}F_{s/d}(\mathbf{q}) \tag{4.22}$$

with

$$F_{s/d}(\mathbf{q}) = \frac{D_{s/d}^2 \cdot [\mathbf{q}]^2}{D_{s/d}^2 + [\mathbf{q}]^2}. \tag{4.23}$$

4.2.2 Yukawa Couplings

The Yukawa couplings $\bar{h}_a(Q)$, $\bar{h}_\rho(Q)$ and $\bar{h}_s(Q)$ are parameterized by means of a linear momentum dependence

$$\bar{h}_{a/\rho/s}(Q) = \frac{|\boldsymbol{\pi}-\mathbf{q}|}{|\boldsymbol{\pi}|}\bar{h}_{a/\rho/s}(0) + \frac{|\mathbf{q}|}{|\boldsymbol{\pi}|}\bar{h}_{a/\rho/s}(\Pi), \tag{4.24}$$

The computation of the flows of $\bar{h}_{a/\rho/s}(0)$ and $\bar{h}_{a/\rho/s}(\Pi)$ will be discussed in Chapter 4.4.1.

For the d-boson, as known from other studies (see, for example, Fig. 7 (b) of [44]), when the four-fermion coupling in the d-wave channel becomes critical it has a sharp peak around zero momentum. This is accounted for by including a Gaussian function which is centered around zero momentum in the definition of $\bar{h}_d$.

$$\bar{h}_d(Q) = \bar{h}_d(0) \exp(-|\mathbf{q}|^2/w_0^2). \qquad (4.25)$$

It has been checked that the results are practically independent of the width w_0^2 of this Gaussian function, as long as it is reasonably peaked. In principle, an ansatz of the form (4.24) might also be used.

4.3 Initial Conditions and Regulators

At the microscopic scale $k = \Lambda$ the flowing action must be equivalent to the microscopic action of the Hubbard model, so the initial value of the four-fermion coupling must correspond to the Hubbard interaction U. The bosonic fields decouple completely at this scale, so the initial values of the Yukawa couplings are

$$\bar{h}_a|_\Lambda = \bar{h}_\rho|_\Lambda = \bar{h}_s|_\Lambda = \bar{h}_d|_\Lambda = 0. \qquad (4.26)$$

For the bosonic mass terms one can take $\bar{m}_{i,\Lambda}^2 = t^2$ and then use units $t = 1$, and $P_{i,\Lambda}(Q) = 0$ for the kinetic terms. The choice $\bar{m}_{i,\Lambda}^2 = t^2$ amounts to an arbitrary choice for the normalization of the bosonic fields, which are introduced as redundant auxiliary fields at the scale $k = \Lambda$, where they do not couple to the electrons. Of course, this changes during the flow, where the bosons are transformed into dynamical composite degrees of freedom, with nonzero Yukawa couplings and a nontrivial momentum dependence of their propagators. The quartic bosonic couplings $\bar{\lambda}_a$, $\bar{\lambda}_d$ and $\bar{\lambda}_{ad}$ vanish on initial scale $k = \Lambda$, the fermionic wave function renormalization is $Z_F(\omega_Q)|_\Lambda = 1$ for all values of the Matsubara frequency ω_Q.

In addition to the truncation for the effective average action, regulator functions for the fermions and bosons have to be specified. Momentum space "optimized" cutoffs [71, 72, 67] are used for both fermions and bosons. The regulator function for fermions is given by

$$R_k^F(Q) = \text{sgn}(\xi(\mathbf{q}))\,(\mathrm{k} - |\xi(\mathbf{q})|)\,\Theta(\mathrm{k} - |\xi(\mathbf{q})|). \qquad (4.27)$$

Its derivative with respect to the scale k, which will often be needed, is given by

$$\partial_k R_k^F(Q) = \text{sgn}(\xi(\mathbf{q}))\,\Theta(\mathrm{k} - |\xi(\mathbf{q})|). \qquad (4.28)$$

Contributions to the flow equations of couplings in which this term occurs are nonzero only for modes having no more than a certain nonzero distance from the Fermi surface. This distance shrinks while k is decreasing, so

4.4 Flow Equations for the Running Couplings

one can describe the renormalization group flow as approaching the Fermi surface step by step while taking into account fluctuations closer and closer to it.

The regulator function for the antiferromagnetic boson is given by

$$R_k^a(Q) = A_a \left(k^2 - F_{a,c/i}(\mathbf{q}, \hat{q})\right) \Theta(k^2 - F_{a,c/i}(\mathbf{q}, \hat{q})) \quad (4.29)$$

allowing for an incommensurability $\hat{q}$ with $F_{a,c/i}$ as defined in Chapter 4.2. For the ρ-boson the same regulator is used, but with $F_{\rho,c/i}$ defined with a shift by the vector π, as discussed in Chapter 4.2. Regulator functions for the Cooper-pair bosons can be chosen of the same form, but no incommensurability needs to be accounted for in these cases, so one can set

$$R_k^{s/d}(Q) = A_{s/d} \cdot (k^2 - F_{s/d}(\mathbf{q})) \Theta(k^2 - F_{s/d}(\mathbf{q})) \quad (4.30)$$

for them.

The derivative of, for instance, the antiferromagnetic regulator with respect to the scale k is given by

$$\partial_k R_k^a(Q) = k A_a \left(2 - \eta_a(1 - F_{a,c/i}(\mathbf{q}, \hat{q}) t^2 / k^2)\right) \Theta(k^2 - F_{a,c/i}(\mathbf{q}, \hat{q})), \quad (4.31)$$

where quantitatively irrelevant renormalization group improvements due to the scale dependence of $F_{a,c/i}$ have been neglected.

While in the case of fermions the last fluctuations taken into account are associated to momenta in the vicinity of the Fermi surface, the last bosonic fluctuations to be integrated out are those which are close to the minima of the kinetic terms. If incommensurate (charge density or antiferromagnetic) fluctuations dominate, there are four such minima, otherwise one. As will become clear in Chapter 5 where phases exhibiting spontaneous breaking are studied, this number has a considerable influence on the strength with which bosonic fluctuation drive the system out of the spontaneously broken regimes in the infrared limit $k \to 0$.

4.4 Flow Equations for the Running Couplings

There exist different ways of extracting the flow equations for the running couplings from the flow equation (3.17) for the flowing action. In this work, two different methods to arrive at these flow equations are applied, the first based on the one-loop corrections to the 1PI vertex functions, the second based on the flow equation for the effective potential. In this section I shall focus on the first of these two approaches, which is used to derive the flow of the running couplings in the symmetric regime. Furthermore, since it involves the computation of the one-loop corrections to the 1PI vertex functions, it produces among others the one-loop correction to the inverse antiferromagnetic propagator Eq. (2.23), which was needed in the mean field treatment of antiferromagnetism described in Chapter 2.5.

As a starting point, the flow equation for the flowing action is written in the form

$$\partial_k \Gamma_k = \frac{1}{2} \text{STr} \, \tilde{\partial}_k \left(\ln(\Gamma_k^{(2)} + R_k) \right) \qquad (4.32)$$

where the shorthand $\tilde{\partial}_k = (\partial_k R_k) \frac{\partial}{\partial R_k}$ has been used.

The basic idea for deriving from this equation the flow equations for the 1PI vertex functions is to derive both sides an appropriate number of times with respect to the fields collected in the collective field $\hat{\chi} = (\hat{\mathbf{a}}, \hat{\rho}, \hat{s}, \hat{s}^*, \hat{d}, \hat{d}^*, \hat{\psi}, \hat{\psi}^*)$ (the "hat" is just for notational convenience in what follows). This is achieved by decomposing $\hat{\chi}$ into a spatially homogeneous background field χ and a fluctuation-dependent part $\delta\hat{\chi}(Q)$ such that

$$\hat{\chi}(Q) = \chi \cdot \delta(Q) + \delta\hat{\chi}(Q). \qquad (4.33)$$

The fermionic fields $\psi^\dagger$ have vanishing expectation values, so one can set $\hat{\psi}^\dagger(Q) = \delta\psi^\dagger(Q)$ in this case. For the bosonic fields, however, the decomposition Eq. (4.33) is nontrivial.

The matrix $\Gamma_k^2 + R_k$, which includes the inverse propagator together with the infrared regulator terms, is now split up into a fluctuation-independent part $\mathcal{P}$, which includes the regulator terms R_k^i and a fluctuation-dependent part $\mathcal{F}$ such that

$$\Gamma_k^2 + R_k = \mathcal{P} + \mathcal{F}. \qquad (4.34)$$

The matrix $\mathcal{P}$ includes the inverse propagators, background fields and cutoff functions. In the symmetric regime it is diagonal in momentum space and therefore easily invertible.

Inserting the decomposition Eq. (4.34) into the Wetterich flow equation (4.32) and expanding in the number of fields around zero fluctuation-dependent part, one obtains

$$\begin{aligned}\partial_k \Gamma_k &= \frac{1}{2} \text{STr} \, \tilde{\partial}_k \left(\ln(\mathcal{P} + \mathcal{F}) \right) \\ &= \frac{1}{2} \text{STr} \, \tilde{\partial}_k \left(\ln \mathcal{P} \right) + \frac{1}{2} \text{STr} \, \tilde{\partial}_k \left(\mathcal{P}^{-1} \mathcal{F} \right) - \frac{1}{4} \text{STr} \, \tilde{\partial}_k \left((\mathcal{P}^{-1} \mathcal{F})^2 \right) \\ &\quad + \frac{1}{6} \text{STr} \, \tilde{\partial}_k \left((\mathcal{P}^{-1} \mathcal{F})^3 \right) - \frac{1}{8} \text{STr} \, \tilde{\partial}_k \left((\mathcal{P}^{-1} \mathcal{F})^4 \right) + \ldots \end{aligned} \qquad (4.35)$$

Following [73], the fluctuation part $\mathcal{F}$ can be split up into parts containing each only one specific combination of types of field,

$$\mathcal{F} = F_{\mathbf{a}} + F_\rho + \ldots + F_\psi + F_{\psi^\dagger} + \ldots + F_{ad} + \ldots . \qquad (4.36)$$

4.4 Flow Equations for the Running Couplings

This decomposition of $\mathcal{F}$ induces a decomposition of $\mathcal{P}^{-1}\mathcal{F}$ such that

$$\mathcal{P}^{-1}\mathcal{F} = N_{\mathbf{a}} + N_\rho + \ldots + N_\psi + N_{\psi^\dagger} + \ldots + N_{ad} + \ldots, \quad (4.37)$$

where $N_{i\ldots} = \mathcal{P}^{-1}F_{i\ldots}$ for all field (multi-) indices $i\ldots$.

The flow equations for the 1PI vertex functions are obtained from Eq. (4.34) by computing the terms on the right hand side up to a certain order in the fields while at first omitting the scale derivative $\tilde{\partial}_k$ and the regulator term R_k so that in accordance with Eq. (3.7) one computes the one-loop correction of the effective action Γ, starting from the classical action S. One-loop corrections to the 1PI vertex functions are obtained by differentiating both sides of Eq. (4.34) (while still neglecting the scale derivative $\tilde{\partial}_k$) with respect to the fluctuation-dependent parts of the components of the field $\hat{\chi}$. In other words, the matrices N_i, as introduced in Eq. (4.37), are compared to those terms in the truncation for Γ which depend on the same types of field i. As an example, the one-loop correction to the inverse antiferromagnetic propagator $\tilde{P}_a(Q)$ is obtained from the term on the right hand side which is quadratic in $\mathbf{a}$ and depends on no other field.

After having computed the one-loop corrections to the 1PI vertex functions, one has to insert the derivative operator $\tilde{\partial}_k$ under the traces occurring in Eq. (4.34), which means introducing it under the frequency, spin and momentum sums (or integrals) of these loop corrections. The flow equations for the running couplings are thus obtained from the loop corrections to the 1PI vertex functions by performing under the loop integral the derivative with respect to the regulator functions and afterwards multiplying with their k-derivatives before carrying out the frequency, spin and momentum summations.

4.4.1 Flow Equations for the Yukawa Couplings

In order to give a comprehensive overview of all one-loop corrections to the 1PI vertex functions, the language of Feynman diagrams is used in what follows. As described in the last section, the flow equations for the running couplings follow from the one-loop corrections to the 1PI vertex functions having an appropriate number of external lines, if one inserts the derivative $\tilde{\partial}_k$ acting onto the IR regulator R_k under the loop integral. In this section the resulting flow equations are discussed, first for the Yukawa couplings, subsequently for the bosonic propagators, the quartic bosonic couplings, and finally for the fermionic wave function renormalization.

Due to the contributions from flowing bosonization, as described in Chapter 3.3, the derivation of the flow equations for the Yukawa couplings is more involved than the derivation of those for the bosonic propagators and quartic couplings. Nevertheless, the derivation of the flow equations for the Yukawa couplings is discussed first because contributions to the flow equations of the bosonic propagators and quartic bosonic couplings are nonzero only if nonzero Yukawa couplings have been generated in the first place. Since according to the choice of initial conditions specified in Chapter 4.2 all Yukawa couplings are equal to zero on initial scale $k = \Lambda$, the flow of all other quantities starts only after nonzero Yukawa couplings have already been generated during the flow.

The flow equations for the Yukawa couplings consist of a direct contribution and an "indirect" contribution resulting from flowing bosonization, see Chapter 3.3. Applying the same line of argument which, for the coupling in the magnetic channel, leads to Eq. (3.24), one obtains for the Yukawa couplings in the four channels taken into account flow equations of the form

$$\partial_k \bar{h}_a(Q) = \partial_k \bar{h}_a\big|_{\mathbf{a}_k}(Q) + \frac{\tilde{P}_{a,k}(Q)}{2\bar{h}_a(Q)} \partial_k \lambda_F^a\big|_{\mathbf{a}_k}(Q),$$

$$\partial_k \bar{h}_\rho(Q) = \partial_k \bar{h}_\rho\big|_{\rho_k}(Q) + \frac{\tilde{P}_{\rho,k}(Q)}{2\bar{h}_\rho(Q)} \partial_k \lambda_F^\rho\big|_{\rho_k}(Q),$$

$$\partial_k \bar{h}_s(Q) = \partial_k \bar{h}_s\big|_{s_k,s_k^*}(Q) + \frac{\tilde{P}_{s,k}(Q)}{2\bar{h}_s(Q)} \partial_k \lambda_F^s\big|_{s_k,s_k^*}(Q),$$

$$\partial_k \bar{h}_d(Q) = \partial_k \bar{h}_d\big|_{d_k,d_k^*}(Q) + \frac{\tilde{P}_{d,k}(Q)}{2\bar{h}_d(Q)} \partial_k \lambda_F^d\big|_{d_k,d_k^*}(Q). \quad (4.38)$$

According to the parameterization for the Yukawa couplings specified in Eqs. (4.24) and (4.25) these equations have to be evaluated at momentum 0 and, for the **a**-, ρ- and s-boson, also at momentum Π. In order to avoid having to divide by the Yukawa couplings which are zero at the beginning of the renormalization flow, it is more convenient to study the flow of their squares. Consequently, the following equations are needed for the Yukawa coupling in the antiferromagnetic channel

$$\partial_k \bar{h}_a^2(0) = \partial_k \bar{h}_a^2\big|_{\mathbf{a}_k}(0) + \tilde{P}_a(0) \partial_k \lambda_F^a\big|_{\mathbf{a}_k}(0),$$
$$\partial_k \bar{h}_a^2(\Pi) = \partial_k \bar{h}_a^2\big|_{\mathbf{a}_k}(\Pi) + \tilde{P}_a(\Pi) \partial_k \lambda_F^a\big|_{\mathbf{a}_k}(\Pi), \quad (4.39)$$

and analogously for $\partial_k \bar{h}_\rho^2(0)$, $\partial_k \bar{h}_\rho^2(\Pi)$, $\partial_k \bar{h}_s^2(0)$, $\partial_k \bar{h}_s^2(\Pi)$ and $\partial_k \bar{h}_d^2(0)$.

The approximation made by computing $\bar{h}_a^2(0)$ according to Eq. (4.39) is most adequate when the loop contributions to $\lambda_F^a(K_1 - K_2)$ are minimal for $K_1 - K_2 = 0$ and maximal for $K_1 = K_2 = \Pi$ or inversely. This is the case whenever either ferromagnetic or *commensurate* antiferromagnetic fluctuations dominate. When *incommensurate* antiferromagnetic fluctuations dominate, the the flow equation for $\bar{h}_a^2(0)$ is taken to be of the form

$$\partial_k \bar{h}_a^2(0) = \partial_k \bar{h}_a^2\big|_{\mathbf{a}_k}(0) + \bar{m}_a^2 \partial_k \lambda_F^a\big|_{\mathbf{a}_k}(0), \quad (4.40)$$

where $\bar{m}_a^2$ is the minimum of the inverse antiferromagnetic propagator which coincides with its value at momenta $\pm \hat{Q}_x = \pm(0, \hat{q}, 0)$ and $\pm \hat{Q}_y = \pm(0, 0, \hat{q})$.

4.4 Flow Equations for the Running Couplings

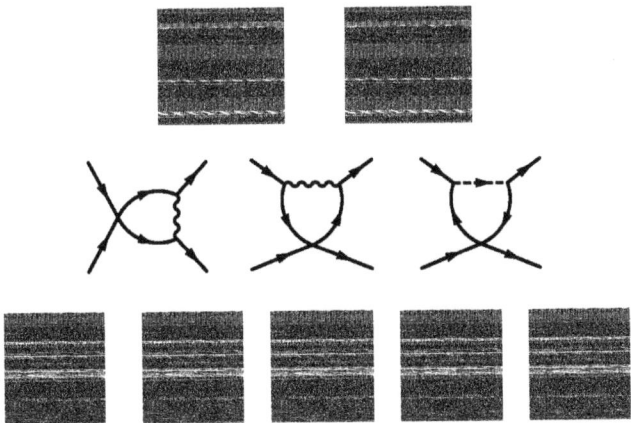

Figure 4.1: 1PI diagrams contributing to the flow of the Yukawa couplings via flowing bosonization.

I shall first discuss those diagrams which contribute to the flow of the Yukawa couplings via flowing bosonization and afterwards those that contribute directly. Those that contribute via flowing bosonization are displayed in Fig. 4.1. There are three types of such diagrams: two diagrams involving only fermionic lines (first line of Fig 4.1), three diagrams having one internal bosonic line (second line of Fig 4.1), and five box diagrams having two internal bosonic lines (third line of Fig 4.1). The first graph in each line of Fig. 4.1 is a particle-particle graph, meaning that one cannot perform a closed loop while following the direction of all arrows involved, all others are particle-hole graphs. As will become clear in what follows, contributions from particle-particle graphs can be absorbed by Cooper pair bosons, contributions from particle-hole graphs by real bosons.

Purely fermionic loops

In order to demonstrate how the contributions to the four-fermion vertex are taken into account via flowing bosonization, I discuss in some detail the case of the purely fermionic loop diagrams shown in the upper line of Fig. 4.1. Their contribution to the effective action is determined from

$$-\frac{1}{4}\text{STr}\left((P^{-1}F)^2\right) \supset -\frac{1}{4}\text{STr}\left(N_{\psi\psi} + N_{\psi\psi^\dagger} + N_{\psi^\dagger\psi^\dagger}\right)^2$$
$$\supset -\frac{1}{4}\text{STr}\left(2N_{\psi\psi}N_{\psi^\dagger\psi^\dagger} + N_{\psi\psi^\dagger}N_{\psi\psi^\dagger}\right), \quad (4.41)$$

where the cyclicity property of the trace has been used and the sign '⊃' denotes projection of the loop corrections on those parts which contain an appropriate number of fields ψ and $\psi^\dagger$.

The resulting loop correction to the effective action is given by

$$\Delta \Gamma_F^F = -\frac{U^2}{2} \sum_{K_1,K_2,K_3,K_4} \sum_P \qquad (4.42)$$
$$\left(\frac{1}{P_F(P)P_F(P+K_2-K_3)} + \frac{1}{P_F(P)P_F(-P+K_1+K_3)} \right)$$
$$\times \delta(K_1 - K_2 + K_3 - K_4) \left[\psi^\dagger(K_1)\psi(K_2)\right] \cdot \left[\psi^\dagger(K_3)\psi(K_4)\right].$$

In order to obtain the corresponding contribution to the fermionic four-point vertex function $\Delta \Gamma_F^{F\,(4)}$, one has to take the fourth functional derivative of $\Delta \Gamma_F^F$ with respect to the fermionic fields. It is given by

$$\Delta \Gamma_F^{F\,(4)}(K_1, K_2, K_3, K_4) = \frac{1}{4} \frac{\delta^4}{\delta \psi_\alpha^*(K_1) \delta \psi_\beta(K_2) \delta \psi_\gamma^*(K_3) \delta \psi_\delta(K_4)} \Delta \Gamma_F^F$$
$$= -\frac{U^2}{4} \sum_P \left(\frac{4 S_{\alpha\gamma;\beta\delta}}{P_F(P)P_F(-P+K_1+K_3)} \right.$$
$$\left. -\frac{\delta_{\alpha\delta}\delta_{\gamma\beta}}{P_F(P)P_F(P+K_2-K_1)} + \frac{\delta_{\alpha\beta}\delta_{\gamma\delta}}{P_F(P)P_F(P+K_2-K_3)} \right), \quad (4.43)$$

where $S_{\alpha\gamma;\beta\delta} = \frac{1}{2}(\delta_{\alpha\beta}\delta_{\gamma\delta} - \delta_{\alpha\delta}\delta_{\gamma\beta})$ denotes the singlet projection. In order to read off from this expression the loop contributions to the four fermion vertex in its different channels, one has to compare the two last lines of Eq. (4.43) to the fourth derivative of the right hand sides of Eqs. (4.6)-(4.9) with respect to the fields ψ, $\psi^\dagger$. This gives the loop corrections to the four-fermion couplings λ_F^a, λ_F^ρ, λ_F^s, λ_F^d introduced there. The second to last line of Eq. (4.43) can be absorbed by the s-boson, the last line by the **a**- and ρ-bosons. Since all terms in Eq. (4.43) depend on only one combination of momenta K_1, K_2 and K_3, no contributions to the coupling in the d-wave channel arise at this stage.

For the last line of Eq. (4.43) it is not immediately seen how it should be distributed onto the **a**- and ρ-bosonic channels. In order to achieve this, one can use the identity (B.11). All terms have now the same structures as those appearing in the fourth functional derivatives of Eqs. (4.6), (4.7) and (4.8). Consequently, one obtains from the purely fermionic loops (first line

4.4 Flow Equations for the Running Couplings

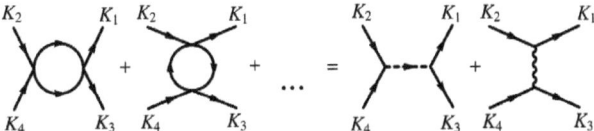

Figure 4.2: Schematic picture of the bosonization of loop contributions to the four-fermion vertex. The terms indicated by the three dots correspond to loop diagrams having internal bosonic lines.

of Fig. 4.1) the following contributions to λ_F^a, λ_F^ρ and λ_F^s:

$$(\Delta\lambda_F^a)^F (K_1 - K_2) = -\frac{U^2}{2} \sum_P \frac{1}{P_F(P)P_F(P + K_2 - K_1)},$$

$$(\Delta\lambda_F^\rho)^F (K_1 - K_2) = -\frac{U^2}{2} \sum_P \frac{1}{P_F(P)P_F(P + K_2 - K_1)}, \quad (4.44)$$

$$(\Delta\lambda_F^s)^F (K_1 + K_3) = \frac{U^2}{2} \sum_P \frac{1}{P_F(P)P_F(-P + K_1 + K_3)}.$$

As described above, one can now extract the flow equations for λ_F^a, λ_F^ρ, λ_F^s from the one-loop expressions (4.44) by replacing P_F by P_F^k (i.e. adding the infrared cutoff R_k^F to the inverse fermionic propagator) and by applying the formal derivative $\tilde{\partial}_k = (\partial_k R_k^F)\partial/\partial R_k^F$ under the summation. For λ_F^a, for example, one obtains

$$\partial_k \lambda_F^a(Q) = \tilde{\partial}_k \Delta\lambda_F^a(Q), \quad (4.45)$$

where the formal derivative $\tilde{\partial}_k$ should be read as acting under the loop summation of terms contained in $\Delta\lambda_F^a(Q)$. Of course $(\Delta\lambda_F^a)^F(Q)$ is only part of the complete loop contribution $\Delta\lambda_F^a(Q)$, namely, the one which arises from the two diagrams shown in the first line of Fig. 4.1. The complete $\Delta\lambda_F^a(Q)$ is obtained if the diagrams shown in Fig. 4.1 are all taken together.

In the partially bosonized approach used here, the fermionic loop contributions to the momentum-dependent four-fermion vertex in Eq. (4.44) are fully accounted for by the exchange of the bosons **a**, ρ and s. This is shown schematically in Fig. 4.2. In the language of boson exchange, the momentum dependence of the coupling in, for instance, the antiferromagnetic channel can be taken into account by the momentum dependence of the expression $\bar{h}_a^2(K_1 - K_2)\tilde{P}_a^{-1}(K_1 - K_2)$. According to the flowing bosonization scheme described in Chapter 3.3, the coupling λ_F^a is kept at zero during the flow, and all contributions to it are taken into account by a corresponding change in the Yukawa couplings $\bar{h}_a^2$. At momentum $Q = 0$, for example, the contribution to the flow of the momentum-dependent Yukawa couplings due to the diagrams in the first line of Fig. 4.1, according to Eq. (4.39), is given by

$$\left(\partial_k \bar{h}_a^2(0)\right)^F = -\frac{U^2}{2}\tilde{P}_a(0)\sum_P \tilde{\partial}_k \frac{1}{P_F^k(P)P_F^k(P-\Pi)}, \quad (4.46)$$

$$\left(\partial_k \bar{h}_\rho^2(0)\right)^F = -\frac{U^2}{2}\tilde{P}_\rho(0)\sum_P \tilde{\partial}_k \frac{1}{P_F^k(P)P_F^k(P)}, \quad (4.47)$$

$$\left(\partial_k \bar{h}_s^2(0)\right)^F = \frac{U^2}{2}\tilde{P}_s(0)\sum_P \tilde{\partial}_k \frac{1}{P_F^k(P)P_F^k(-P)}. \quad (4.48)$$

For the s-boson $\tilde{P}_s(0)$ equals the mass term $\bar{m}_s^2$, for the **a**-boson $\tilde{P}_a(0)$ equals the mass term $\bar{m}_a^2$ if commensurate antiferromagnetic fluctuations dominate over incommensurate ones.

At the level of Eqs. (4.46)-(4.48), the one-loop perturbative result for the momentum-dependent four-fermion vertex is completely described in terms of boson exchange. In the purely fermionic flows [22, 23, 24, 25, 26, 27, 28, 29] the constant coupling U would be replaced by the full momentum- and k-dependent four-fermion vertex. According to the partially bosonized approach used here, in contrast, where only a constant four-fermion coupling U is kept, this renormalization group improvement is generated by the diagrams involving internal bosonic lines, shown in the second and third lines of Fig. 4.1, and by the direct contributions shown in Fig. 4.3. It is at this level where the parameterization of the Yukawa couplings and inverse boson propagators specified in Chapter 4.2 as well as the restriction to a certain number of bosons starts to matter.

Diagrams with one internal bosonic line

The momentum dependence of the four-fermion vertex arising from the diagrams involving boson exchange (those in the second and third lines of Fig. 4.1) is much more complicated than the simple form (4.44). The decision of how to distribute these contributions onto the different boson exchange channels is therefore nontrivial. Here an approximation is adopted in which the momentum-dependence of the four-fermion couplings λ_F^a, λ_F^ρ, λ_F^s, λ_F^d is identified with the dependence of the diagrams in Figs. 4.1 and 4.3 on the so-called transfer momentum. This momentum is defined as the difference between the momenta attached to the two fermionic propagators in each diagram. According to this prescription particle-hole diagrams are absorbed by the real bosons and particle-particle diagrams by the complex Cooper pair bosons.

All diagrams are evaluated at external momenta $L = (\pi T, \pi, 0)$ and $L' = (\pi T, 0, \pi)$ so that, in accordance with Eqs. (4.39) and (4.40), flow equations for the Yukawa couplings are obtained at transfer momenta $0 = (0,0,0)$

4.4 Flow Equations for the Running Couplings

and $\Pi = (0, \pi, \pi)$. For small values of $|\mu|$ and $|t'|$, the spatial components l and l' of the momenta L and L' are close to the Fermi surface. Since for not so large values of $|\mu|$ and $|t'|$ the density of states is comparatively large at l and l' and sometimes even divergent, this choice of external momenta for the evaluation of diagrams is likely to be adequate for the most important scattering processes. Where more than one combination of external momenta $\pm L$ and $\pm L'$ is compatible with the condition that the transfer momentum is either 0 or Π, the average over these is taken.

While the contributions to the Yukawa couplings in Eqs. (4.46) and (4.48) are proportional to U^2 (and not to any Yukawa coupling) and therefore present already for large k, the diagrams shown in Fig. 4.3 and in the second and third lines of Fig. 4.1 start to have an influence on the flow of the Yukawa couplings only after nonzero Yukawa couplings have been generated from Eqs. (4.46)-(4.48) in the first place. In perturbation theory, they would correspond to higher order effects $\sim U^3$ and U^4. (Perturbatively, every Yukawa coupling counts as U.)

The corrections to the effective action due to particle-particle diagrams ("pp" denotes "particle-particle") with one internal real bosonic line read

$$\Delta \Gamma^{a,\rho}_{F,pp} = -\frac{U}{2} \sum_{K_1,K_2,K_3,K_4} \sum_{P}$$
$$\left(\frac{\sigma^j_{\alpha\delta}\sigma^j_{\gamma\beta}}{P_F(P)P_F(-P+K_1+K_3)} \left(\frac{\bar{h}_a^2(P+K_1-\Pi)}{\tilde{P}_a(P+K_1-\Pi)} + \frac{\bar{h}_a^2(P+K_2-\Pi)}{\tilde{P}_a(P+K_2-\Pi)} \right) \right.$$
$$\left. + \frac{\delta_{\alpha\delta}\delta_{\gamma\beta}}{P_F(P)P_F(-P+K_1+K_3)} \left(\frac{\bar{h}_\rho^2(P+K_1)}{\tilde{P}_\rho(P+K_1)} + \frac{\bar{h}_\rho^2(P+K_2)}{\tilde{P}_\rho(P+K_2)} \right) \right)$$
$$\times \delta(K_1 - K_2 + K_3 - K_4) \psi^\dagger_\alpha(K_1)\psi_\beta(K_2)\psi^\dagger_\gamma(K_3)\psi_\delta(K_4) . \quad (4.49)$$

Using the identities (B.10) and (B.11), the projection of the fourth derivative with respect to the fermionic fields onto the singlet channel (hence the index "s") is computed to be

$$\Delta \Gamma^{(4),a,\rho}_{F,pp,s}(K_1, K_2, K_3, K_4) = -\frac{U}{4} \sum_{P}$$
$$\left(\frac{3}{P_F(P)P_F(-P+K_1+K_3)} \left(\frac{\bar{h}_a^2(P+K_1-\Pi)}{\tilde{P}_a(P+K_1-\Pi)} + \frac{\bar{h}_a^2(P+K_2-\Pi)}{\tilde{P}_a(P+K_2-\Pi)} \right.\right.$$
$$\left. + \frac{\bar{h}_a^2(P+K_3-\Pi)}{\tilde{P}_a(P+K_3-\Pi)} + \frac{\bar{h}_a^2(P+K_4-\Pi)}{\tilde{P}_a(P+K_4-\Pi)} \right)$$
$$- \frac{1}{P_F(P)P_F(-P+K_1+K_3)} \left(\frac{\bar{h}_\rho^2(P+K_1)}{\tilde{P}_\rho(P+K_1)} + \frac{\bar{h}_\rho^2(P+K_2)}{\tilde{P}_\rho(P+K_2)} \right.$$
$$\left.\left. + \frac{\bar{h}_\rho^2(P+K_3)}{\tilde{P}_\rho(P+K_3)} + \frac{\bar{h}_\rho^2(P+K_4)}{\tilde{P}_\rho(P+K_4)} \right) \right) . \quad (4.50)$$

Evaluated at external momenta $K_i = \pm L^{(\prime)}$ so that $K_1+K_3 = K_2+K_4 = 0$, this yields the following contribution to the flow of $\bar{h}_s^2(0)$,

$$\left(\partial_k \bar{h}_s^2(0)^2\right)^{a,\rho} = \bar{m}_s^2 \frac{U}{2} \sum_P \tilde{\partial}_k \left(\frac{3\bar{h}_a^2(P-L)}{P_F^k(P)P_F^k(-P)\tilde{P}_a^k(P-L)} \right. $$
$$\left. - \frac{\bar{h}_\rho^2(P-L)}{P_F^k(P)P_F^k(-P)\tilde{P}_\rho^k(P-L)} \right), \quad (4.51)$$

where it has been used that it is irrelevant whether L or L' occurs in the summation over P. In the analogous expression for $\bar{h}_s^2(\Pi)$ an additional shift by the vector Π has to be introduced in the argument of one in each pair of fermionic propagators P_F^k in the denominators of Eq. (4.51).

Particle-hole graphs (index "ph") with one internal bosonic line can have either a real or a Cooper pair boson internal line. Those with a real boson internal line are given by

$$\Delta\Gamma_{F,ph}^{F,a} = \frac{U}{4} \sum_{K_1,K_2,K_3,K_4} \sum_P \frac{1}{P_F(P)P_F(P-K_1+K_2)}$$
$$\left(\frac{\bar{h}_a^2(P+K_1-\Pi)}{\tilde{P}_a(P+K_1-\Pi)} + \frac{\bar{h}_a^2(P+K_2-\Pi)}{\tilde{P}_a(P+K_2-\Pi)} \right) \left(3\delta_{\alpha\beta}\delta_{\gamma\delta} + \sigma_{\alpha\beta}^j \sigma_{\gamma\delta}^j \right),$$
$$\times \delta(K_1-K_2+K_3-K_4) \psi_\alpha^\dagger(K_1)\psi_\beta(K_2)\psi_\gamma^\dagger(K_3)\psi_\delta(K_4), \quad (4.52)$$

$$\Delta\Gamma_{F,ph}^{F,\rho} = \frac{U}{4} \sum_{K_1,K_2,K_3,K_4} \sum_P \frac{1}{P_F(P)P_F(P-K_1+K_2)}$$
$$\left(\frac{\bar{h}_\rho^2(P+K_1)}{\tilde{P}_\rho(P+K_1)} + \frac{\bar{h}_\rho^2(P+K_2)}{\tilde{P}_\rho(P+K_2)} \right) \left(\delta_{\alpha\beta}\delta_{\gamma\delta} - \sigma_{\alpha\beta}^j \sigma_{\gamma\delta}^j \right)$$
$$\times \delta(K_1-K_2+K_3-K_4) \psi_\alpha^\dagger(K_1)\psi_\beta(K_2)\psi_\gamma^\dagger(K_3)\psi_\delta(K_4), \quad (4.53)$$

those with a Cooper pair boson internal line by

$$\Delta\Gamma_{F,ph}^{F,s} = -U \sum_{K_1,K_2,K_3,K_4} \sum_P \frac{1}{P_F(P)P_F(P-K_1+K_2)}$$
$$\left(\frac{\bar{h}_s^2(P+K_1)}{\tilde{P}_s(P+K_1)} + \frac{\bar{h}_s^2(P+K_2)}{\tilde{P}_s(P+K_2)} \right) \left(\delta_{\alpha\beta}\delta_{\gamma\delta} + \sigma_{\alpha\beta}^j \sigma_{\gamma\delta}^j \right)$$
$$\times \delta(K_1-K_2+K_3-K_4) \psi_\alpha^\dagger(K_1)\psi_\beta(K_2)\psi_\gamma^\dagger(K_3)\psi_\delta(K_4), \quad (4.54)$$

$$\Delta\Gamma_{F,ph}^{F,d} = -U \sum_{K_1,K_2,K_3,K_4} \sum_P \frac{f_d(P/2-K_1)f_d(P/2+K_2)}{P_F(P)P_F(P-K_1+K_2)}$$
$$\left(\frac{\bar{h}_d^2(P+K_1)}{\tilde{P}_d(P+K_1)} + \frac{\bar{h}_d^2(P+K_2)}{\tilde{P}_d(P+K_2)} \right) \left(\delta_{\alpha\beta}\delta_{\gamma\delta} + \sigma_{\alpha\beta}^j \sigma_{\gamma\delta}^j \right)$$
$$\times \delta(K_1-K_2+K_3-K_4) \psi_\alpha^\dagger(K_1)\psi_\beta(K_2)\psi_\gamma^\dagger(K_3)\psi_\delta(K_4). \quad (4.55)$$

4.4 Flow Equations for the Running Couplings

Performing the derivative with respect to the fermionic fields, evaluating at external momenta $K_i = \pm L^{(\prime)}$, and inserting the scale derivative $\tilde{\partial}_k$, the corresponding contributions to $\bar{h}_a^2(0)$ and $\bar{h}_\rho^2(0)$ are given by

$$\left(\partial_k \bar{h}_a^2(0)\right)^{apsd} = -\bar{m}_a^2 \frac{U}{2} \sum_P \tilde{\partial}_k \frac{1}{P_F^k(P) P_F^k(P+\Pi)}$$

$$\left(\frac{\bar{h}_a^2(P+L)}{\tilde{P}_a^k(P+L)} - \frac{\bar{h}_\rho^2(P+L)}{\tilde{P}_\rho^k(P+L)} \right. \quad\quad (4.56)$$

$$\left. -4\frac{\bar{h}_s^2(P+L)}{\tilde{P}_s^k(P+L)} - 4\frac{\bar{h}_d^2(P+L)}{\tilde{P}_d^k(P+L)} f_d((P-L)/2) f_d((P+\Pi+L')/2) \right)$$

and

$$\left(\partial_k \bar{h}_\rho^2(0)\right)^{apsd} = -\tilde{P}_\rho(0) \frac{U}{2} \sum_P \tilde{\partial}_k \frac{1}{P_F^k(P) P_F^k(P)}$$

$$\left(3\frac{\bar{h}_a^2(P+L)}{\tilde{P}_a^k(P+L)} + \frac{\bar{h}_\rho^2(P+L)}{\tilde{P}_\rho^k(P+L)} \right. \quad\quad (4.57)$$

$$\left. -4\frac{\bar{h}_s^2(P+L)}{\tilde{P}_s^k(P+L)} - 4\frac{\bar{h}_d^2(P+L)}{\tilde{P}_d^k(P+L)} f_d((P-L)/2) f_d((P+L')/2) \right) .$$

The analogous contributions to the flows of $\bar{h}_a^2(\Pi)$ and $\bar{h}_\rho^2(\Pi)$ have the same form. The single refinement is that there is no "+Π" in the argument of the second fermionic propagator and in the second d-wave form factor in the equation for $\bar{h}_a^2(\Pi)$ and an additional "+Π" in the argument of the second fermionic propagator and in the second d-wave form factor in the equation for $\bar{h}_\rho^2(\Pi)$.

Box diagrams: Generation of a coupling in the d-wave channel

The detailed evaluation of the box diagram contributions to the four-fermion vertex, which are given in the third line of Fig. 4.1, can be found in Appendix C of the present work. Here I focus on the generation of a coupling in the d-wave channel through the particle-particle box graph with antiferromagnetic internal lines.

Since the direct contribution to the d-wave Yukawa coupling $\bar{h}_d$ is proportional to this coupling itself, it is nonzero only if $\bar{h}_d$ is nonzero in the first place. As the initial value of $\bar{h}_d$ at the scale $k = \Lambda$ is zero, $\bar{h}_d$ has to grow from zero due to other contributions, namely the first diagram in the lower line of Fig. 4.1, which is the only particle-particle box graph. The contribution to the four-fermion vertex in this channel from this graph is extracted by means of the prescription

$$\Delta\lambda_F^d(0) = \frac{1}{2}\{\Delta\Gamma_{F,s}^{(4),pp}(L,L,-L,-L) - \Delta\Gamma_{F,s}^{(4),pp}(L,L',-L,-L')\}, \quad (4.58)$$

where the subscript "s" denotes the singlet and the superscript "pp" the particle-particle part of the four-point vertex. The momentum vectors L and L' are defined as in Appendix A. For a motivation of this prescription as a way of extracting the d-wave coupling see [34], for further details see Appendix C, in particular Eqs. (C.3) and (C.4). The resulting contribution to the flow of the d-wave Yukawa coupling from the particle-particle box diagram with two internal antiferromagnetic lines is given by

$$\left(\partial_k \bar{h}_d(0)^2\right)^{aa} = \bar{m}_d^2 \frac{9}{16} \sum_{\epsilon=\pm 1} \sum_P \tilde{\partial}_k \quad (4.59)$$
$$\left(\frac{\bar{h}_a^2(P+L)}{P_F^k(P)P_F^k(-P)\tilde{P}_a^k(P+L)} \left(\frac{\bar{h}_a^2(P+\epsilon L)}{\tilde{P}_a^k(P+\epsilon L)} - \frac{\bar{h}_a^2(P+\epsilon L')}{\tilde{P}_a^k(P+\epsilon L')}\right)\right).$$

The contribution from the same diagram to the s-wave superconducting channel is obtained by adding, instead of subtracting, the two terms on the right-hand side of Eq. (4.58) (and analogously in Eq. (4.59)). The s- and d-wave superconducting channels of the four-fermion coupling can be described as those parts of its singlet particle-particle contribution which are symmetric (s-wave) and antisymmetric (d-wave) with respect to rotations by 90° of the outgoing electrons while keeping the incoming electrons fixed.

After a Yukawa coupling in the d-wave channel has been generated in accordance with Eq. (4.59) it is further enhanced through its direct contributions. The discussion now turns to these.

Direct contributions

Having discussed the contributions to the flow of the Yukawa couplings which arise from the flowing bosonization of contributions to the four-fermion vertex, I now come to the direct contributions, which are shown graphically in Fig. 4.3.

The corrections to the effective action corresponding to the diagrams without an internal bosonic line, shown in the first line of Fig. 4.3, are given

4.4 Flow Equations for the Running Couplings

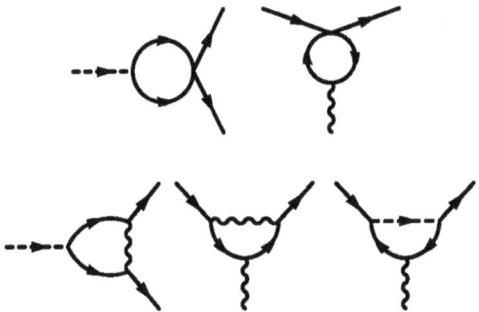

Figure 4.3: 1PI diagrams which directly contribute to the flow of the Yukawa couplings.

by

$$\Delta\Gamma^Y_{a,F} = U \sum_{K,Q_1,Q_2} \sum_P \frac{\bar{h}_a(K)}{P_F(P)P_F(P-K-\Pi)} \quad (4.60)$$
$$\times \delta\mathbf{a}(K) \cdot [\psi^\dagger(Q_1)\boldsymbol{\sigma}\psi(Q_2)]\delta(K - Q_1 + Q_2 - \Pi),$$

$$\Delta\Gamma^Y_{\rho,F} = -U \sum_{K,Q_1,Q_2} \sum_P \frac{\bar{h}_\rho(K)}{P_F(P)P_F(P-K)} \quad (4.61)$$
$$\times \delta\rho(K) \cdot [\psi^\dagger(Q_1)\psi(Q_2)]\delta(K - Q_1 + Q_2),$$

for the real and

$$\Delta\Gamma^Y_{s,F} = U \sum_{K,Q_1,Q_2} \sum_P \frac{\bar{h}_s(K)}{P_F(P)P_F(-P+K)} \quad (4.62)$$
$$\left(\delta s^*(K) \cdot [\psi^T(Q_1)\epsilon\psi(Q_2)] - \delta s(K) \cdot [\psi^\dagger(Q_1)\epsilon\psi^*(Q_2)]\right)\delta(K - Q_1 - Q_2),$$

$$\Delta\Gamma^Y_{d,F} = U \sum_{K,Q_1,Q_2} \sum_P \frac{\bar{h}_d(K) f_d(P+(Q_1-Q_2)/2)}{P_F(P)P_F(-P+K)} \quad (4.63)$$
$$\left(\delta d^*(K) \cdot [\psi^T(Q_1)\epsilon\psi(Q_2)] - \delta d(K) \cdot [\psi^\dagger(Q_1)\epsilon\psi^*(Q_2)]\right)\delta(K - Q_1 - Q_2)$$

for the complex bosons.

Deriving with respect to the fields, evaluating at external boson momentum $K = 0$, the insertion of $\tilde{\partial}_k$ yields the contributions

$$\left(\bar{h}_a(0)\right)^{F,\mathrm{dir}} = -\bar{h}_a(0)U\sum_P \tilde{\partial}_k \frac{1}{P_F^k(P)P_F^k(P+\Pi)}, \quad (4.64)$$

$$\left(\bar{h}_\rho(0)\right)^{F,\mathrm{dir}} = \bar{h}_\rho(0)U\sum_P \tilde{\partial}_k \frac{1}{P_F^k(P)P_F^k(P)}, \quad (4.65)$$

$$\left(\bar{h}_s(0)\right)^{F,\mathrm{dir}} = -\bar{h}_s(0)U\sum_P \tilde{\partial}_k \frac{1}{P_F^k(P)P_F^k(-P)}, \quad (4.66)$$

$$\left(\bar{h}_d(0)\right)^{F,\mathrm{dir}} = 0. \quad (4.67)$$

For bosonic momentum $K = \Pi$ one proceeds in exactly the same way, which has the consequence that the "$+\Pi$" in one of the inverse fermionic propagators vanishes in the first and is additionally introduced in the other lines. The contribution to $\bar{h}_d$ is zero since the integration over spatial momenta vanishes due to the $d_{x^2-y^2}$-symmetry of the d-wave form factor.

At the beginning of the renormalization flow the growth of the Yukawa couplings in the antiferromagnetic and charge density channels according to Eqs. (4.46) and (4.47) is the same. They start to differ, however, as soon as the direct contributions to the Yukawa couplings shown in the first line of Fig. 4.3 become important. They contribute positively to the coupling in the magnetic channel but negatively to the couplings in the charge density and superconducting s-wave channels. This explains why in the parameter regimes investigated among the three Yukawa couplings $\bar{h}_a, \bar{h}_\rho, \bar{h}_s$ the dominating one is $\bar{h}_a$, although in accordance with Eqs. (4.46)-(4.48) all three are generated equally at early stages of the flow. Due to the comparatively large Yukawa coupling $\bar{h}_a$ the mass term $\bar{m}_a^2$ is driven fastest toward zero. It thus becomes understandable why the charge density and s-wave superconducting channels do not become critical in the range of parameters investigated.

Further direct contributions to the Yukawa-couplings, shown in the second line of Fig. 4.3, have an internal bosonic line. The associated one-loop corrections to the effective action read as follows:

$$\Delta\Gamma_{a,F}^Y = \sum_{K,Q_1,Q_2}\sum_P \quad (4.68)$$

$$\left(\frac{\bar{h}_a(K)}{P_F(Q_1 - K + P)P_F(Q_1 + P - \Pi)} \cdot \left(\frac{\bar{h}_a^2(P)}{\tilde{P}_a(P)} - \frac{\bar{h}_\rho^2(P)}{\tilde{P}_\rho(P)}\right.\right.$$
$$\left.\left. - \frac{4\bar{h}_s^2(P)}{\tilde{P}_s(P)} - \frac{4\bar{h}_d^2(P)f_d(Q_1 - \Pi + P/2)f_d(Q_1 - K + P/2)}{\tilde{P}_d(P)}\right)\right)$$
$$\times \delta\mathbf{a}(K)\left[\psi^\dagger(Q_1)\boldsymbol{\sigma}\psi(Q_2)\right]\delta(K - Q_1 + Q_2 - \Pi),$$

4.4 Flow Equations for the Running Couplings

$$\Delta\Gamma^Y_{\rho,F} = \sum_{K,Q_1,Q_2}\sum_P \tag{4.69}$$

$$\left(\frac{\bar{h}_\rho(K)}{P_F(Q_1-K+P)P_F(Q_1+P)}\cdot\left(-\frac{3\bar{h}_a^2(P)}{\tilde{P}_a(P)}-\frac{\bar{h}_\rho^2(P)}{\tilde{P}_\rho(P)}\right.\right.$$

$$\left.\left.+\frac{4\bar{h}_s^2(P)}{\tilde{P}_s(P)}+\frac{4\bar{h}_d^2(P)f_d(Q_1-\Pi+P/2)f_d(Q_1-K+P/2)}{\tilde{P}_d(P)}\right)\right)$$

$$\times\delta\rho(K)\left[\psi^\dagger(Q_1)\boldsymbol{\sigma}\psi(Q_2)\right]\delta(K-Q_1+Q_2-\Pi),$$

and

$$\Delta\Gamma^Y_{s,F} = \sum_{K,Q_1,Q_2}\sum_P \tag{4.70}$$

$$\left(\frac{\bar{h}_s(K))}{P_F(Q_1-K+P)P_F(Q_1+P)}\left(-3\frac{\bar{h}_a^2(P)}{\tilde{P}_a(P)}+\frac{\bar{h}_\rho^2}{\tilde{P}_\rho(P)}\right)\right)$$

$$\left(\delta s^*(K)\cdot\left[\psi^T(Q_1)\epsilon\psi(Q_2)\right]-\delta s(K)\cdot\left[\psi^\dagger(Q_1)\epsilon\psi^*(Q_2)\right]\right)\delta(K-Q_1-Q_2),$$

$$\Delta\Gamma^Y_{d,F} = \sum_{K,Q_1,Q_2}\sum_P \tag{4.71}$$

$$\left(\frac{\bar{h}_d(K)f_d(P+Q_1-K/2)}{P_F(Q_1-K+P)P_F(Q_1+P)}\left(-3\frac{\bar{h}_a^2(P)}{\tilde{P}_a(P)}+\frac{\bar{h}_\rho^2}{\tilde{P}_\rho(P)}\right)\right)$$

$$\left(\delta d^*(K)\cdot\left[\psi^T(Q_1)\epsilon\psi(Q_2)\right]-\delta d(K)\cdot\left[\psi^\dagger(Q_1)\epsilon\psi^*(Q_2)\right]\right)\delta(K-Q_1-Q_2).$$

To extract from these loop corrections to the effective action the corresponding contributions to the flow equations of the Yukawa couplings, the derivative with respect to the fields is taken and evaluated at external boson momentum $K=0$ and fermionic momenta $\pm L^{(\prime)}$. Inserting $\tilde{\partial}_k$ under the loop integral one obtains

$$(\partial_k\bar{h}_a)^{a\rho sd,\text{dir}}(0) = \bar{h}_a(0)\sum_P\tilde{\partial}_k\frac{1}{P_F^k(P)P_F^k(P+\Pi)} \tag{4.72}$$

$$\times\left(-\frac{\bar{h}_a^2(P+L)}{\tilde{P}_a^k(P+L)}+\frac{\bar{h}_\rho^2(P+L)}{\tilde{P}_\rho^k(P+L)}+4\frac{\bar{h}_s^2(P+L)}{\tilde{P}_s^k(P+L)}\right.$$

$$\left.+4\frac{\bar{h}_d^2(P+L)f_d((P-L)/2)f_d((P+\Pi+L')/2)}{\tilde{P}_d^k(P+L)}\right)$$

and, for $\bar{h}_\rho(0)$,

$$\left(\partial_k \bar{h}_\rho\right)^{a\rho sd,\text{dir}}(0) = \bar{h}_\rho(0) \sum_P \tilde{\partial}_k \frac{1}{P_F^k(P)P_F^k(P+L)} \qquad (4.73)$$

$$\times \left(3\frac{\bar{h}_a^2(P+L)}{\tilde{P}_a^k(P+L)} + \frac{\bar{h}_\rho^2(P+L)}{\tilde{P}_\rho^k(P+L)} - 4\frac{\bar{h}_s^2(P+L)}{\tilde{P}_s^k(P+L)} \right.$$

$$\left. -4\frac{\bar{h}_d^2(P+L)f_d((P-L)/2)f_d((P+L')/2)}{\tilde{P}_d^k(P+L)}\right).$$

In the parameter regimes studied, those contributions in Eqs. (4.72) and (4.73) that have a minus sign tend to contribute positively to the growth of the Yukawa couplings, the others negatively. This means that growing $\bar{h}_a$ enhances its own growth, whereas growing $\bar{h}_\rho$ blocks its own growth. On the other hand, the growth of $\bar{h}_a$ is slowed down by contributions from the s-wave superconducting channel (see both Eq. (4.56) and Eq. (4.72)). Intuitively, the superconducting channels describe an effective attraction between fermions that has a tendency to attenuate the dominant repulsion which in turn may lead to antiferromagnetic order. Taking into account the s-wave superconducting boson is crucial in order not to obtain critical temperatures that are too large by a factor of approximately 2.[1]

For the direct contributions to the flow of $\bar{h}_s(0)$ and $\bar{h}_d(0)$ involving a bosonic propagator one obtains:

$$\left(\partial_k \bar{h}_s(0)\right)^{a\rho,\text{dir}} = \bar{h}_s(0) \sum_P \tilde{\partial}_k \frac{1}{P_F(P)P_F(-P)} \qquad (4.74)$$

$$\times \left(-\frac{3\bar{h}_a^2(P+L)}{\tilde{P}_a(P+L)} + \frac{\bar{h}_\rho^2(P+L)}{\tilde{P}_\rho(P+L)}\right)$$

$$\left(\partial_k \bar{h}_d(0)\right)^{a\rho,\text{dir}} = \bar{h}_d(0) \sum_P \tilde{\partial}_k \frac{f_d(P)}{P_F(P)P_F(-P)} \qquad (4.75)$$

$$\times \left(-\frac{3\bar{h}_a^2(P+L)}{\tilde{P}_a(P+L)} + \frac{\bar{h}_\rho^2(P+L)}{\tilde{P}_\rho(P+L)}\right)$$

The presence of the d-wave form factor $f_d(P)$ in the contribution to $\partial_k \bar{h}_d(0)$ is crucial. Once a coupling in the d-wave channel has been generated through

---

[1] Earlier studies using the framework employed in this work (for instance [31, 32, 34, 35]) did not obtain (pseudo-) critical temperatures too large by a factor of approximately 2 even though no s-wave superconducting boson was taken into account there. The main reason why the (pseudo-) critical temperatures for antiferromagnetism were not grossly overestimated in these works is that the momenta chosen for the evaluation of diagrams contributing to $\bar{h}_a$ were not close to the Fermi surface, which led to an inadequate estimate of these contributions. This second inadequacy essentially compensated for the neglect of the s-boson, so that plausible-looking results for the antiferromagnetic (pseudo-) critical temperatures were obtained.

4.4 Flow Equations for the Running Couplings

the particle-particle box diagram (first in the third line of Fig 4.1), it is further enhanced due to the direct contribution computed in Eq. (4.75). Since this expression, which is itself proportional to the Yukawa coupling $\bar{h}_d(0)$, contributes positively to the flow of $\bar{h}_d$ due to the d-wave form factor, it can lead to a growth of this coupling without bounds, i.e. lead to an an instability in the d-wave channel. This instability will be the result of antiferromagnetic spin fluctuations so that the results of the present work, finding a d-wave instability due to this contribution, support the idea, proposed and defended in [4, 5, 6, 7, 8, 9, 10], that antiferromagnetic spin fluctuations are responsible for d-wave superconductivity in the two-dimensional Hubbard model (and maybe also in the cuprates insofar as the Hubbard model serves as a guide to the relevant cuprate physics).

That the particle-particle graph in the second line of Fig. 4.3 holds a key role in the emergence of the d-wave superconducting instability arising from antiferromagnetic fluctuations is mirrored by the fact that this diagram has the same momentum structure as right hand side of the BCS gap equation. In the presence of an interaction which in momentum space is maximal around the (π, π)-points—a condition which is fulfilled when antiferromagnetic spin fluctuations dominate—the gap solving this equation exhibits d-wave symmetry.

4.4.2 Bosonic Propagators

The derivation of the flow equations for the momentum-dependent bosonic propagators starts from the projection of the one-loop corrections to the effective action onto the parts which are quadratic in the bosonic fields. For the antiferromagnetic boson, for instance, one has to project onto the part which is quadratic in the field **a** such that

$$\Delta\Gamma \supset -\frac{1}{4}\mathrm{STr}\left(\mathbf{N}_a^2\right) = \sum_Q \Delta\Gamma_{aa}^{(2)}(Q)\cdot\frac{1}{2}\mathbf{a}(-Q)\mathbf{a}(Q), \quad (4.76)$$

where $\Delta\Gamma_{aa}^{(2)}(Q)$ is to be identified with the one-loop correction to the inverse antiferromagnetic propagator $\Delta P_a(Q)$. It is obtained from $\Delta\Gamma$ by performing the second derivative with respect to **a**,

$$\Delta P_a(Q) = \frac{\overrightarrow{\delta}}{\delta \mathbf{a}^T(-Q)}\Delta\Gamma\frac{\overleftarrow{\delta}}{\delta \mathbf{a}(Q)}\bigg|_{\mathbf{a}=0}. \quad (4.77)$$

If the Yukawa coupling $\bar{h}_a$ were assumed to be independent of frequency and momentum, the fermionic loop contribution $\Delta\left(P_a(Q)\right)^F$ would give the result presented in the mean field analysis of Chapter 2.5, see Eq. (2.23). According to the parameterization of the Yukawa couplings introduced in Chapter 4.2, however, $\bar{h}_a$ depends on spatial momenta so that

$$\Delta\left(P_a(Q)\right)^F = 2\bar{h}_a^2(Q)\sum_P \frac{1}{P_F(Q+P+\Pi)P_F(P)}. \quad (4.78)$$

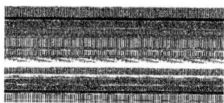 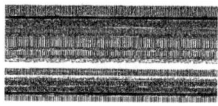

Figure 4.4: Fermionic loop diagrams contributing to the flow of bosonic propagators. Wiggly lines denote real bosons (particle-hole channels), dashed lines complex bosons (Cooper pair channels).

Diagrammatically, this expression is presented as the first graph in Fig. 4.4. For the ρ-boson one obtains an analogous result, simply replacing the index a by ρ and omitting the vector Π in the first inverse fermionic propagator, whereas for the d-boson one gets

$$\Delta\left(P_d(Q)\right)^F = -4\bar{h}_d^2(Q) \sum_P \frac{f_d(P)^2}{P_F(Q+P)P_F(-P)} \tag{4.79}$$

and analogously for the s-boson after replacing the index d by s and the d-wave form factor f_d by 1. The diagrammatic representation of Eq. (4.78) is the second graph in Fig.4.4.

Inserting the derivative $\tilde{\partial}_k$, one obtains from Eqs. (4.78) and (4.79) the following fermionic contributions to the flow equations for the inverse bosonic propagators,

$$\begin{aligned}
\left(\partial_k \tilde{P}_a(Q)\right)^F &= 2\bar{h}_a^2(Q) \sum_P \tilde{\partial}_k \frac{1}{P_F^k(P)P_F^k(P+\Pi+Q)} \\
\left(\partial_k \tilde{P}_\rho(Q)\right)^F &= 2\bar{h}_\rho^2(\hat{Q}) \sum_P \tilde{\partial}_k \frac{1}{P_F^k(P)P_F^k(P+Q)} \\
\left(\partial_k \tilde{P}_s(Q)\right)^F &= -4\bar{h}_s^2(0) \sum_P \tilde{\partial}_k \frac{1}{P_F^k(-P)P_F^k(P+Q)} \\
\left(\partial_k P_d(Q)\right)^F &= -4\bar{h}_d^2(Q) \sum_P \tilde{\partial}_k \frac{f_d(\mathbf{p})^2}{P_F^k(-P)P_F^k(P+Q)}.
\end{aligned} \tag{4.80}$$

For the **a**-and d-boson further contributions to the inverse propagators are taken into account. They are proportional to the quartic bosonic couplings $\bar{\lambda}_a$, $\bar{\lambda}_d$ and $\bar{\lambda}_{ad}$ and independent of the external momentum Q:

$$\Delta\left(P_a(Q)\right)^B = \frac{5}{2}\bar{\lambda}_a \sum_P \frac{1}{\tilde{P}_a(P)} + \bar{\lambda}_{ad} \sum_P \frac{1}{\tilde{P}_d(P)} \tag{4.81}$$

and

$$\Delta\left(P_d(Q)\right)^B = -2\bar{\lambda}_d \sum_P \frac{1}{\tilde{P}_d(P)} + \frac{3}{2}\bar{\lambda}_{ad} \sum_P \frac{1}{\tilde{P}_a(P)}. \tag{4.82}$$

4.4 Flow Equations for the Running Couplings

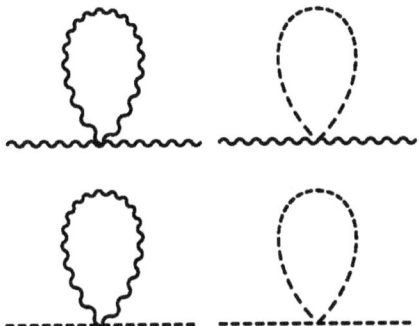

Figure 4.5: Diagrams involving the quartic bosonic couplings $\bar{\lambda}_a$, $\bar{\lambda}_d$ and $\bar{\lambda}_{ad}$ whose the scale derivatives contribute to the flow of the antiferromagnetic and d-wave superconducting propagators. Scale derivatives of the diagrams in the first line contribute to the flow of the antiferromagnetic, those in the second line to that of the d-wave superconducting propagator. Wiggly lines denote antiferromagnetic, dashed lines superconducting propagators.

The bosonic mass terms are defined as the minima of the inverse propagators $\tilde{P}_i(Q)$. Taking into account both the fermionic contributions Eqs. (4.78) and (4.79) and the bosonic contributions (4.81) and (4.82), the flow of the antiferromagnetic and d-wave superconducting mass terms is given by

$$\partial_k \bar{m}_a^2 = 2\bar{h}_a^2 \sum_P \tilde{\partial}_k \frac{1}{P_F^k(P) P_F^k(P + \Pi + \hat{Q})} \quad (4.83)$$

$$- \sum_P \tilde{\partial}_k \left(\frac{5}{2} \frac{\bar{\lambda}_a}{\tilde{P}_a^k(P)} + \frac{\bar{\lambda}_{ad}}{\tilde{P}_d^k(P)} \right)$$

and

$$\partial_k \bar{m}_d^2 = -4\bar{h}_d^2 \sum_P \tilde{\partial}_k \frac{f_d(\mathbf{p})^2}{P_F^k(P) P_F^k(-P)} \quad (4.84)$$

$$- \sum_P \tilde{\partial}_k \left(2 \frac{\bar{\lambda}_d}{P_d^k(P) + \bar{m}_d^2} + \frac{3}{2} \frac{\bar{\lambda}_{ad}}{P_a^k(P) + \bar{m}_a^2} \right).$$

The vector $\hat{Q} = \hat{Q}_x = (0, \hat{q}, 0)$ (or $\hat{Q} = \hat{Q}_y = (0, 0, \hat{q})$) occurring in the second fermionic propagator in Eq. (4.83) contains the incommensurability which, as explained in Chapter 2.5, characterizes the magnetic and charge density fluctuations in a certain range of parameters.

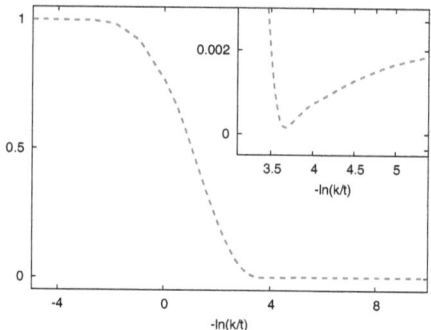

Figure 4.6: Flow of the antiferromagnetic mass term $\bar{m}_a^2$ for $U/t = 3$, $t'/t = -0.l$, $\mu/t = -0.77$ and $T/t = 0.0185$. The inset shows a detail of the flow, namely the point where $\bar{m}_a^2$ reaches its minimum, followed by an increase, which is caused by to the bosonic contributions in the second line of Eq. (4.83).

In order to determine the value of the incommensurability $\hat{q}$, the right hand side of Eq. (4.83) is computed for a large number of vectors $\hat{Q}$ corresponding to different values of $\hat{q}$. The value of $\hat{q}$ for which, at a given scale k, the mass term $\bar{m}_a^2$ is minimal is accepted as the true value of the incommensurability and the associated value of $\bar{m}_a^2$ as the true antiferromagnetic mass term.

In the symmetric regime, the fermionic contributions involving the Yukawa couplings decrease the mass terms during the renormalization flow whereas the bosonic contributions, proportional to the quartic couplings $\bar{\lambda}_a$, $\bar{\lambda}_d$ and $\bar{\lambda}_{ad}$ (if it is positive), tend to increase them. The closer the mass terms approach zero, the more important the bosonic fluctuations become. As the charge density and s-wave superconducting channels never become critical in the range of parameters investigated, the contributions to their mass terms are assumed to be small and therefore neglected. Once either the **a**- or **d**- boson mass term becomes close to zero, bosonic fluctuations become more important, and the terms in the second lines of Eqs. (4.83) and (4.84) may prevent the mass term from actually reaching zero. Fig. 4.6 shows this in an example where the bosonic contribution proportional to $\bar{\lambda}_a$ inverts the direction of the flow of the mass term $\bar{m}_a^2$ so that it remains nonzero for $k \to 0$.

Whenever some bosonic mass term $\bar{m}_i^2$ becomes zero during the flow—despite the influence of bosonic fluctuations—the truncation for the effective potential is changed from the form of Eq. (4.15) to one which is appropriate for the phases with broken symmetry, see Eq. (5.1) in Chapter 5.1. A

4.4 Flow Equations for the Running Couplings

negative quadratic term in the effective potential indicates local order, since at a given coarse graining scale k the effective average action evaluated at constant field has a minimum for a nonzero value of the boson field. The largest temperature where at fixed values of U, t', μ a given mass term $\bar{m}_i^2$ vanishes during the flow is called the pseudocritical temperature T_{pc} for this type of order. It can also be described as the largest temperature where short-range order sets in. At this temperature the effective momentum-dependent four-fermion coupling $\bar{h}_i^2(Q)/\bar{m}_i^2$ diverges in the channel where $\bar{m}_i^2$ hits zero. If the order persists for k reaching a macroscopic scale, the model exhibits effectively spontaneous symmetry breaking, associated in the Hubbard model in the parameter regimes studied to (either commensurate or incommensurate) antiferromagnetism or d-wave superconductivity. The true critical temperature T_c is defined as the largest temperature for which local order persists up to some physical scale k_{ph} corresponding to the inverse size of a macroscopic sample, see [31, 33]. In order to determine the true critical temperature for either **a**- or d-type of order, it is therefore necessary to switch to the truncation in which either α_0 or δ_0 (or both) are nonzero. How this is done will be discussed in Chapter 5.

The flows of the factors Z_i and A_i (with $i =$ **a**, ρ, s, d) which occur in the parameterization of the bosonic propagators are obtained as difference quotients from the flows of $\tilde{P}_i(Q)$, evaluated at appropriate values of Q. The flows of Z_a and A_a, for instance, are computed from the difference quotients displayed in Eq. (4.21) and analogously for the other bosons.

It is convenient to introduce the anomalous dimensions η_i, which are defined through

$$\eta_i = -k\partial_k \ln A_i. \tag{4.85}$$

They are a measure of how quickly the gradient coefficients A_i change with the scale k, so they can be determined from the flows of these.

4.4.3 Quartic Bosonic Couplings

The flow of the quartic bosonic couplings $\bar{\lambda}_a$, $\bar{\lambda}_d$, and $\bar{\lambda}_{ad}$ is crucial for the long-range physics of the system in the symmetry-broken regimes, which is dominated by bosonic fluctuations. In order to obtain the appropriate starting values for these couplings in the symmetry-broken regimes, it is necessary to consider their flows already in the symmetric regime. If commensurate antiferromagnetic fluctuations dominate, the flow equation for

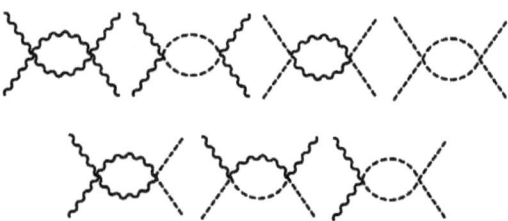

Figure 4.7: Diagrams involving the quartic bosonic couplings $\bar{\lambda}_a$, $\bar{\lambda}_d$ and $\bar{\lambda}_{ad}$ whose the scale derivatives contribute to the flow of these couplings themselves. Scale derivatives of the first two diagrams in the first line contribute to the flow of $\bar{\lambda}_a$, those of the third and fourth diagram in the first line to the flow of $\bar{\lambda}_d$, and those in the second line to the flow of $\bar{\lambda}_{ad}$. Wiggly lines denote antiferromagnetic, dashed lines superconducting propagators.

the antiferromagnetic quartic coupling $\bar{\lambda}_a$ is given by

$$\partial_k \bar{\lambda}_a = \Delta \dot{\Gamma}_a^{(4)}(0,0,0,0)$$
$$= 4\bar{h}_a^4(0) \sum_P \tilde{\partial}_k \frac{1}{\left(P_F^k(P)P_F^k(P+\Pi)\right)^2} \quad (4.86)$$
$$- \sum_P \tilde{\partial}_k \left(\frac{11}{2} \frac{\bar{\lambda}_a^2}{(P_a^k(P)+\bar{m}_a^2)^2} + \frac{\bar{\lambda}_{ad}^2}{(P_d^k(P)+\bar{m}_d^2)^2} \right),$$

where $\Delta \Gamma_a^{(4)}$ denotes the one-loop contribution to the bosonic four-point function, obtained as the fourth functional derivative of the flowing action with respect to the field **a**, and the dot ˙ indicates the insertion of $\tilde{\partial}_k$ under the measure of the loop integral implicit in $\Delta \Gamma_a^{(4)}$. Where incommensurate fluctuations dominate over commensurate ones the flow equation (4.86) for $\bar{\lambda}_a$ has to be modified yielding

$$\partial_k \bar{\lambda}_a = \frac{1}{2} \left(\Delta \dot{\Gamma}_a^{(4)}(\hat{Q}_x, -\hat{Q}_x, \hat{Q}_x, -\hat{Q}_x) \right.$$
$$\left. + \Delta \dot{\Gamma}_a^{(4)}(\hat{Q}_x, -\hat{Q}_x, \hat{Q}_y, -\hat{Q}_y) \right). \quad (4.87)$$

For the quartic coupling $\bar{\lambda}_d$ of the d-boson one has the flow equation

$$\partial_k \bar{\lambda}_d(0) = 16\bar{h}_d^4 \sum_P \tilde{\partial}_k \frac{f_d(\mathbf{p})^4}{\left(P_F^k(P)P_F^k(-P)\right)^2} \quad (4.88)$$
$$- \sum_P \tilde{\partial}_k \left(5 \frac{\bar{\lambda}_d^2}{(P_d^k(P)+\bar{m}_d^2)^2} + \frac{3}{2} \frac{\bar{\lambda}_{ad}^2}{(P_a^k(P)+\bar{m}_a^2)^2} \right),$$

4.5 Numerical Results

which leads to very large values of $\bar{\lambda}_d$ during the flow.

The flow equation for the quartic coupling $\bar{\lambda}_{ad}$ describing the mutual interaction between the **a**- and *d*-boson is given by

$$\partial_k \bar{\lambda}_{ad} = 8\bar{h}_a^2(0)\bar{h}_d^2(0) \sum_P \tilde{\partial}_k \left(\frac{-2f_d(\mathbf{p})^2}{\left(P_F^k(P)\right)^2 P_F^k(-P)P_F^k(P+\Pi)} \right.$$
$$\left. + \frac{f_d(\mathbf{p})f_d(\mathbf{p}+\pi)}{P_F^k(P)P_F^k(-P)P_F^k(P+\Pi)P_F^k(-P+\Pi)} \right)$$
$$- \sum_P \tilde{\partial}_k \left(\frac{5}{2} \frac{\bar{\lambda}_a \bar{\lambda}_{ad}}{\left(P_a^k(P) + \bar{m}_a^2\right)^2} + 2 \frac{\bar{\lambda}_d \bar{\lambda}_{ad}}{\left(P_d^k(P) + \bar{m}_d^2\right)^2} \right.$$
$$\left. + 2 \frac{\bar{\lambda}_{ad}^2}{\left(P_a^k(P) + \bar{m}_a^2\right)\left(P_d^k(P) + \bar{m}_d^2\right)} \right). \tag{4.89}$$

Graphical representations of the diagrams from which the bosonic contributions to the flow of $\bar{\lambda}_a$, $\bar{\lambda}_d$ and $\bar{\lambda}_{ad}$ are obtained as scale derivatives are given in Fig. 4.7.

4.4.4 Fermionic Wave Function Renormalization

The flow equation for the fermionic wave function renormalization factor $Z_F = Z_F(\pm \pi)$ is obtained from the one-loop correction to the fermionic propagator at the lowest two Matsubara modes $\pm \pi T$. Here the formula

$$\partial_k Z_F = \frac{1}{2\pi i} \left(\Delta \dot{\Gamma}_F^{(2)}(\pi T, \mathbf{q}_F) - \Delta \dot{\Gamma}_F^{(2)}(-\pi T, \mathbf{q}_F) \right) \tag{4.90}$$

is used where the subscript "F" and the superscript "(2)" in $\Delta \dot{\Gamma}_F^{(2)}$ indicate that the derivative with respect to the fermionic fields ψ and $\psi^\dagger$ has to be taken. The dot $\cdot$ again indicates the insertion of $\tilde{\partial}_k$ under the measure of the loop integral implicit in $\Delta \Gamma_F^{(2)}$. Due to the ansatz (4.4) for the fermionic propagator, where Z_F is approximated by a single constant, some choice has to be made for the Fermi momentum $\mathbf{q}_F$ appearing on the right hand side of Eq. (4.90). As has been checked, the increase of Z_F during the flow is generally stronger for $\mathbf{q}_F$ close to the points $(0, \pm \pi)$ and $(\pm \pi, 0)$ than for $\mathbf{q}_F$ close to the axes where $q_x = \pm q_y$, but the precise choice does not matter for the semi-quantitative features of the phase diagram. For the results displayed in the figures $\mathbf{q}_F$ has been set to $(0, \pi)$.

4.5 Numerical Results

In this section numerical results for the renormalization group analysis of the symmetric regime are presented and discussed.

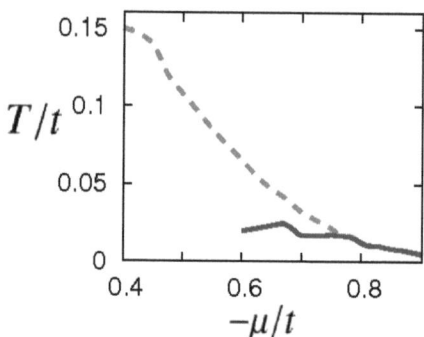

Figure 4.8: Pseudocritical temperatures for antiferromagnetism (dashed) and d-wave superconductivity (solid) at $t'/t = -0.1$ and $U/t = 3$.

Fig. 4.8 shows the pseudocritical temperatures for antiferromagnetic and d-wave superconducting order as a function of the chemical potential μ at $t'/t = -0.1$ and $U/t = 3$, i. e. the highest temperature, depending on μ, for which a given mass term reaches zero during the flow. Antiferromagnetic order is always incommensurate in the range of parameters depicted in Fig. 5.5. Superconducting order sets in only at rather large values of $-\mu$ and only at rather low temperatures compared to the results presented in [36]. The main reason for this seems to be the quartic coupling $\bar{\lambda}_d$, which was not taken into account in [36] but becomes very large during the flow and has a strong tendency to prevent the d-wave superconducting mass term from reaching zero. In order to obtain the true critical temperatures, where magnetic or superconducting order persists up to some macroscopic scale k_{ph}, one has to employ a truncation which allows for nonzero order parameters, i. e. spontaneous symmetry breaking. An analysis of the spontaneously broken regimes will be presented in the following chapter.[2]

In the remaining part of this chapter the qualitative features of the flow of the running couplings in the symmetric regime are discussed, where either antiferromagnetism or d-wave superconductivity is the dominant instability.

The flow of the bosonic mass terms and Yukawa couplings is shown in the upper panels of Fig. 4.9 in the regime where antiferromagnetism is the dom-

---

[2] A nonzero antiferromagnetic order parameter must already be be taken into account to compute the pseudocritical temperatures for d-wave superconductivity where the d-wave pseudocritical curve lies below that for antiferromagnetism in Fig. 4.8. For the truncation used see also Chapter 5.1.

4.5 Numerical Results

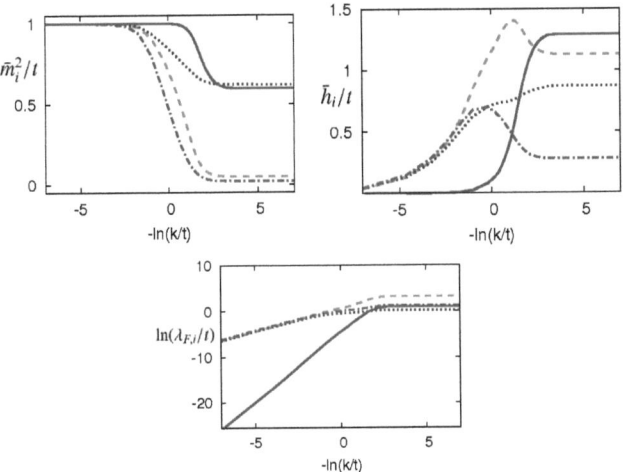

Figure 4.9: Flow of the bosonic mass terms $\bar{m}_a^2$, $\bar{m}_\rho^2$, $\bar{m}_s^2$ and $\bar{m}_d^2$ (upper left panel) and the Yukawa couplings $\bar{h}_a(0)$, $\bar{h}_\rho(\Pi)$, $\bar{h}_s(0)$ and $\bar{h}_d(0)$ (upper right panel). The lower panel shows a logarithmic plot of the effective fermionic four-point couplings $\lambda_{F,i}$ where $\lambda_{F,a} = \bar{h}_a^2(0)/\bar{m}_a^2$, $\lambda_{F,\rho} = \bar{h}_\rho^2(\Pi)/\bar{m}_\rho^2$, $\lambda_{F,s} = \bar{h}_s^2(0)/\bar{m}_s^2$ and $\lambda_{F,d} = \bar{h}_d^2(0)/\bar{m}_d^2$. The lines for all three panels are dashed for the antiferromagnetic boson, solid for the d-wave superconducting boson, dotted for the charge density wave boson, and dashed-dotted for the s-wave superconducting boson. All dimensionful quantities are in units of t. Parameters chosen are $U/t = 3$, $t'/t = -0.1$, $\mu/t = -0.6$ and $T/t = 0.07$, where the system is always in the symmetric regime.

inant instability. Since the s-wave superconducting mass term falls slightly below the antiferromagnetic mass term and the Yukawa coupling in the d-wave channel $\bar{h}_d$ rises above the Yukawa coupling in the antiferromagnetic channel, one has to look at the ratios $\bar{h}_i^2/\bar{m}_i^2$ in order to see that the coupling in the antiferromagnetic channel is actually the dominant one. This is shown in Fig. 4.9, lower panel, where one can see that for the given choice of parameters the antiferromagnetic coupling is more strongly enhanced than the couplings in the s- and d-wave superconducting channels. The coupling in the charge density channel grows least of all four.

In Fig. 4.10 the flow of the bosonic mass terms, Yukawa couplings and effective fermionic four-point couplings is displayed for a combination of parameters where the coupling in the d-wave superconducting channel is the dominant one. Although this coupling is the smallest on high scales of the

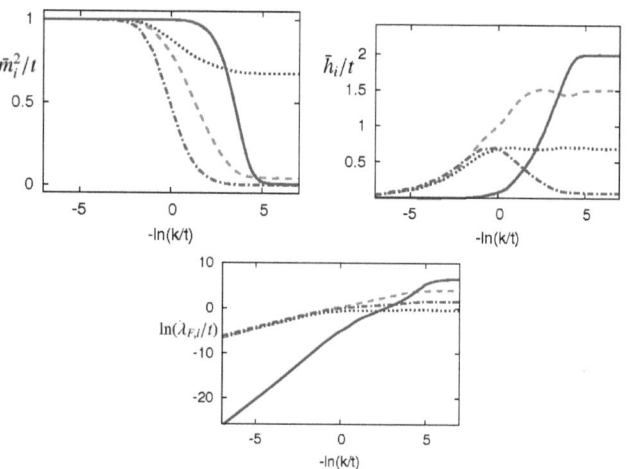

Figure 4.10: Same as Fig. 4.9, now for $U/t = 3$, $t'/t = -0.1$, $\mu/t = -0.83$ and $T/t = 0.011$.

flow by several orders of magnitude, it is strongly enhanced during the flow due to antiferromagnetic fluctuations, as discussed in connection with Eq. (4.75). At temperatures slightly lower than in Fig. 4.10 the mass term $\bar{m}_d^2$ reaches zero and the d-wave coupling diverges at a nonzero renormalization scale $k = k_{\text{SSB}}$.

The flow of the Z- and A-factors used in the parameterization of the a- and d-boson propagators is displayed in the upper panels of Fig. 4.11. The lower panel shows the fermionic wave function renormalization factor $Z_F(\pi T)$, which start its flow from 1 and grows by some fraction for which the increase by 20% in Fig. 4.11 is representative.

The left panel of Fig. 4.12 shows the flow of the quartic bosonic couplings $\bar{\lambda}_a$, $\bar{\lambda}_d$ and $\bar{\lambda}_{ad}$ for the set of parameters also used in Fig. 4.9. Although antiferromagnetism is the dominant instability for this choice of parameters, the quartic coupling $\bar{\lambda}_a$ (dashed curve) is only comparatively weakly enhanced during the flow. For smaller values of $-t'$ and not so close to half filling it may even turn negative during the flow so that the effective potential, according to the truncation (4.15), is no longer bounded from below so that the truncation is no longer adequate and has to be replaced by a more extended one. A negative value of $\bar{\lambda}_a$ may either indicate a tendency towards a first order antiferromagnetic phase transition, but it may also result from

4.5 Numerical Results

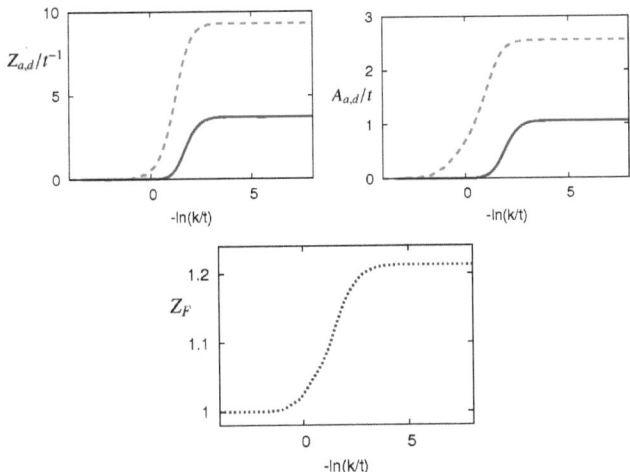

Figure 4.11: Upper left panel: Flow of the bosonic wave function renormalization factors Z_a (dashed) and Z_d (solid). Upper right panel: Flow of the gradient coefficients A_a (dashed) and A_d (solid). Lower panel: Flow of the fermionic wave function renormalization factor $Z_F(\pi T)$. All curves are for the symmetric regime at $U/t = 3$, $t' = -0.1$, $\mu/t = -0.6$ and $T/t = 0.07$.

a more general inadequacy of the parameterization of the effective potential as a polynomial in the fields in the given range of parameters. To avoid these difficulties, which do not arise at larger values of $-\mu$ (see the dashed curve in the right panel of Fig. 4.12), in this work the focus lies on values of the parameters t' and μ for which $\bar{\lambda}_a$ is non-negative on all scales.

While the coupling $\bar{\lambda}_a$ stays rather small during the flow and mostly has only a mild influence on the flow of the antiferromagnetic mass term in SYM, the quartic coupling $\bar{\lambda}_d$ can grow very large. Already for the parameters used in the left panel of Fig. 4.12 where the d-wave channel is far from critical $\bar{\lambda}_d$ (solid curve) is substantially more enhanced than the quartic coupling $\bar{\lambda}_a$, and even much more so in the range of parameters where d-wave superconductivity is the dominant instability, see the right panel of Fig. 4.12 where $\bar{\lambda}_d$ is displayed after division by ten. As already remarked, the eminent growth of $\bar{\lambda}_d$ during the renormalization flow is chiefly responsible for the fact that the transition to d-wave superconductivity occurs only at rather large values of $-\mu$ as compared to the results in [36] where no quartic bosonic couplings were taken into account.

Functional Renormalization for the Symmetric Regime

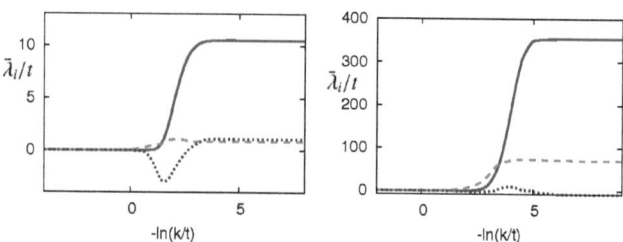

Figure 4.12: Left panel: Flow of the (unrenormalized) quartic bosonic couplings $\bar{\lambda}_a$ (dashed), $\bar{\lambda}_d$ (solid) and $\bar{\lambda}_{ad}$ (dotted) in SYM for $U/t = 3$, $t'/t = -0.1$, $\mu/t = -0.6$ and $T/t = 0.07$. Right panel: The same for $\mu/t = -0.83$ and $T/t = 0.011$, with $\bar{\lambda}_d$ multiplied by 0.1.

The quartic coupling $\bar{\lambda}_{ad}$, which describes the direct interaction between the **a**- and *d*- boson can change its sign from positive to negative, or inversely, during the renormalization flow, see Fig. 4.12 (dotted curves). If it is positive, it enhances the mass terms $\bar{m}_a^2$ and $\bar{m}_d^2$, otherwise it decreases them like the fermionic contributions to their flow.

Chapter 5

Functional Renormalization for the Spontaneously Broken Regimes

In the present chapter the renormalization group analysis started in the previous chapter is extended to the regimes with spontaneously broken symmetries where either one or both of the field expectation values α_0 and δ_0 are nonzero. The regime where only α_0 is nonzero is denoted by "SSBa", the regime where only δ_0 is nonzero is denoted by "SSBd", and the regime where both are nonzero is denoted by "SSBad". The fact that one finds a regime where at a certain scale k of the renormalization flow both order parameters are nonzero does not imply that there is a coexistence phase in which both antiferromagnetic and d-wave superconducting order are present on a scale k_{ph} corresponding to the (inverse) size of a realistic sample. One result from the present investigation is that the region of coexistence of antiferromagnetism and d-wave superconductivity in μ-T-space, if it exists at all, is much less extended than one might have expected from the results for the symmetric regime (see Fig. 4.8).

In Chapter 5.1, the truncation for the symmetry broken regimes is presented. Afterwards, in Chapter 5.2, the flow equations for the running couplings are derived from the flow equation for the local effective potential and the connection to the $O(2)$- and $O(3)$-symmetric linear σ-models is discussed. Chapter 5.3 turns to the results obtained for the temperature dependence of the order parameters, for the phase diagram and for the flow of the running couplings in the symmetry broken regimes.

5.1 Truncation and Approximations

In the spontaneously broken regimes the ρ- and s-bosons are dropped from the truncation since they are expected to have only a very small influence on the running of the field expectation values α_0 and δ_0. In SSBad the minimum of the effective potential occurs at nonzero values of the fields **a** and d. In this case, the effective potential is expanded around its minimum located at (α_0, δ_0):

$$\sum_X U(\mathbf{a}, d) = \frac{1}{2} \sum_{Q_1, Q_2, Q_3, Q_4} \delta(Q_1 + Q_2 + Q_3 + Q_4)$$
$$(\bar{\lambda}_a(\alpha(Q_1, Q_2) - \alpha_0 \delta(Q_1) \delta(Q_2))$$
$$\times (\alpha(Q_3, Q_4) - \alpha_0 \delta(Q_3) \delta(Q_4))$$
$$+ \bar{\lambda}_d(\delta(Q_1, Q_2) - \delta_0 \delta(Q_1) \delta(Q_2)) \quad (5.1)$$
$$(\delta(Q_3, Q_4) - \delta_0 \delta(Q_3) \delta(Q_4))$$
$$+ 2\bar{\lambda}_{ad}(\alpha(Q_1, Q_2) - \alpha_0 \delta(Q_1) \delta(Q_2))$$
$$(\delta(Q_3, Q_4) - \delta_0 \delta(Q_3) \delta(Q_4))).$$

In the regimes $SSBa$ and $SSBd$, where only either α_0 or δ_0 is nonzero, the order parameter corresponding to the symmetry which is unbroken remains zero and the corresponding mass term is kept in the truncation for the effective potential.

In the spontaneously broken regimes, the scale-dependences of the Yukawa couplings $\bar{h}_a$ and $\bar{h}_d$ are neglected, keeping their values fixed at those they have at the scale k_{SSB} where the first mass term becomes zero: $\bar{h}_a(Q)|_k \equiv \bar{h}_a(Q)|_{k_{SSB}}$ and $\bar{h}_d(Q)|_k \equiv \bar{h}_d(Q)|_{k_{SSB}}$ for $k < k_{SSB}$. This approximation, which is made mainly for computational reasons, is expected to have only a minor impact on the flow of the field expectation values in the SSB-regimes at least at low scales k, where it is dominated by long-range bosonic fluctuations more than by interactions between fermions and bosons. As a further simpliciation, the incommensurability $\hat{q}$ is neglected in the SSB-regimes, which would otherwise have to be included in the truncation (5.1) for the effective potential. Although including the incommensurability might influence the flow of the antiferromagnetic order parameter at intermediate scales, its flow at low scales would remain essentially unaffected,as it is mainly determined by the number of Goldstone bosons. In addition to the continuous symmetry associated to the antiferromagnetic order parameter, incommensurate antiferromagnetic order breaks the symmetry of rotations of the lattice by $\pi/2$. However, no Goldstone bosons are associated to the spontaneous breakdown of a discrete symmetry such as this one. The spontaneous breaking of lattice translation invariance thus should not result in major changes of the flow. Setting the incommensurability to zero in the spontaneously broken regimes will presumably leave the universal aspects of the flow of the running couplings in the SSB regimes intact.

The renormalization flow enters the spontaneously broken regimes only at rather low scales, so in all nonzero contributions to the flow of the running couplings the inverse propagators are evaluated close to their minima. Consequently, the spatial momentum part of $P_a(Q)$ can be expanded to quadratic order around its minimum, whose location, assuming the incom-

5.2 Flow of the Effective Potential

mensurability $\hat{q}$ to be zero, is given by $Q = 0$, yielding

$$P_a(Q) = Z_a \omega_Q^2 + A_a[\mathbf{q}]^2 \qquad (5.2)$$

for the region close to $Q = 0$.

Furthermore, since in the spontaneously broken regimes one normally has $k \ll T$, only the lowest bosonic Matsubara mode contributes and the dimensionality of the problem is effectively reduced from $2+1$ to 2, a mechanism which is known as "dimensional reduction". The values of Z_a and Z_d are no longer important and need not be considered any more. The range of temperatures considered is confined to the region $T > T_{min} = 4 \cdot 10^{-3} t$.

To study the universal properties of the flow of the running couplings in the infrared it is convenient to introduce dimensionless (renormalized) quantities $\tilde{\alpha}$, $\tilde{\delta}$, κ_a, κ_d, m_a^2, m_d^2, λ_a, λ_d and λ_{ad} defined by

$$\begin{aligned}
\tilde{\alpha} &= \frac{t^2 A_a}{T}\alpha, & \tilde{\delta} &= \frac{t^2 A_d}{T}\delta, \\
\kappa_a &= \frac{t^2 A_a}{T}\alpha_0, & \kappa_d &= \frac{t^2 A_d}{T}\delta_0, \\
m_a^2 &= \frac{1}{k^2 A_a}\bar{m}_a^2, & m_d^2 &= \frac{1}{k^2 A_d}\bar{m}_d^2, \\
\lambda_a &= \frac{T}{t^2 k^2 A_a^2}\bar{\lambda}_a, & \lambda_d &= \frac{T}{t^2 k^2 A_d^2}\bar{\lambda}_d, \\
\lambda_{ad} &= \frac{T}{t^2 k^2 A_a A_d}\bar{\lambda}_{ad}.
\end{aligned} \qquad (5.3)$$

5.2 Flow of the Effective Potential

In order to obtain the flow equations for the field expectation values and quartic couplings in the SSB-phases, one can derive them from the flow equation for the local effective potential $U(\alpha, \delta)$ which is given by

$$\partial_k U(\alpha, \delta) = \frac{1}{2}\text{STr}\tilde{\partial}_k \ln \mathcal{P}[\alpha, \delta]. \qquad (5.4)$$

Here $\mathcal{P}[\alpha, \delta]$ is the fluctuation-independent part of the cutoff-dependent full inverse propagator $\Gamma_k^{(2)}[\alpha, \delta] + R_k$ including the regulator term, see Eq. (4.34). Within the present truncation, the right hand side of the flow equation for the effective potential (5.4) can be decomposed into a fermionic and a bosonic contribution.

$$\partial_k U(\alpha, \delta) = (\partial_k U(\alpha, \delta))^F + (\partial_k U(\alpha, \delta))^B. \qquad (5.5)$$

The bosonic part can be written as

$$(\partial_k U(\alpha, \delta))^B = \frac{1}{2}\sum_{P,i,j} \tilde{\partial}_k \ln\left[P_i(P)\delta_{i,j} + \hat{M}_{i,j}^2(\alpha, \delta) + R_i^k(P)\delta_{i,j}\right], \qquad (5.6)$$

where $P_i(P) = P_a(P)$ and $R_i^k(P) = R_a^k(P)$ for $i = 1, 2, 3$, and $P_i(P) = P_d(P)$ and $R_i^k(P) = R_d^k(P)$ for $i = 4, 5$, respectively. The matrix $\hat{M}_{i,j}^2(\alpha, \delta)$, which has to be diagonalized, has entries

$$\hat{M}_{i,j}^2(\alpha, \delta) = \begin{cases} \bar{\lambda}_a(3\alpha - \alpha_0) + \bar{\lambda}_{ad}(\delta - \delta_0) & \text{if } i = j = 1, \\ \bar{\lambda}_a(\alpha - \alpha_0) + \bar{\lambda}_{ad}(\delta - \delta_0) & \text{if } i = j = 2, 3, \\ \bar{\lambda}_d(3\delta - \delta_0) + \bar{\lambda}_{ad}(\alpha - \alpha_0) & \text{if } i = j = 4, \\ \bar{\lambda}_d(\delta - \delta_0) + \bar{\lambda}_{ad}(\alpha - \alpha_0) & \text{if } i = j = 5, \\ \frac{1}{2}\bar{\lambda}_{ad}\sqrt{\alpha\delta} & \text{if } i = 1 \text{ and } j = 4, \\ \frac{1}{2}\bar{\lambda}_{ad}\sqrt{\alpha\delta} & \text{if } i = 4 \text{ and } j = 1, \\ 0 & \text{otherwise}. \end{cases} \quad (5.7)$$

The first and fourth lines and columns of the matrix $\hat{M}_{i,j}^2(\alpha, \delta)$ are associated to the radial, the others to the Goldstone modes. The radial modes of the two bosons are coupled to each other through the coupling $\bar{\lambda}_{ad}$ whereas the Goldstone modes remain uncoupled. The form Eq. (5.7) for the matrix $\hat{M}_{i,j}^2(\alpha, \delta)$ is adequate only in SSBad where the minimum of the effective potential $U(\alpha, \delta)$ occurs at nonzero values α_0, δ_0 of both field expectation values α and δ. If one of the symmetries associated to the **a**- or d-boson remains unbroken, the entries associated to this boson have to be replaced by contributions including its mass term. In the presence of an unbroken **a**-symmetry, the first three diagonal entries of $\hat{M}_{i,j}^2(\alpha, \delta)$ have to be replaced by $\bar{m}_a^2 + 3\bar{\lambda}_a\alpha$ (for $i = 1$) and $\bar{m}_a^2 + \bar{\lambda}_a\alpha$ (for $i = 2, 3$). If the d-wave superconducting symmetry remains unbroken, the fourth and fifth diagonal entries of $\hat{M}_{i,j}^2(\alpha, \delta)$ have to be replaced by $\bar{m}_d^2 + 3\bar{\lambda}_d\delta$ and $\bar{m}_d^2 + \bar{\lambda}_d\delta$.

The fermionic contribution $(\partial_k U(\alpha, \delta))^F$ to the flow of the effective potential is obtained from the one-loop correction to the effective potential given by

$$(\Delta U)^F = -\frac{1}{2}\text{Tr}\ln \mathcal{P}_F, \quad (5.8)$$

where the sum in the trace is over fermionic indices only. If the antiferromagnetism is commensurate, one can carry out the Matsubara sum analytically, obtaining (see Section VI in [31])

$$(\Delta U)^F = -T \int_\mathbf{p} \frac{d^2p}{(2\pi)^2} \sum_{\epsilon=\{\pm 1\}} \ln \cosh\left(\frac{\Theta_\epsilon}{2T}\right) \quad (5.9)$$

where

$$\Theta_\epsilon = \left[\left(\frac{1}{2}(\xi_\mathbf{p} + \xi_{\mathbf{p}+\boldsymbol{\pi}}) + \epsilon\sqrt{\frac{1}{4}(\xi_\mathbf{p} - \xi_{\mathbf{p}+\boldsymbol{\pi}})^2 + 2\bar{h}_a^2\alpha_0}\right)^2 + 4\bar{h}_d^2 f_d(\mathbf{p})^2 \delta_0\right]^{1/2}. \quad (5.10)$$

5.2 Flow of the Effective Potential

By the help of Eqs. (5.6) and (5.8) the flow equations for the quartic couplings $\bar{\lambda}_a$, $\bar{\lambda}_d$ and $\bar{\lambda}_{ad}$ are obtained by appropriate derivatives with respect to the fields α and δ on both sides of Eq. (5.4),

$$
\begin{aligned}
\partial_k \bar{\lambda}_a &= \frac{d^2}{d\alpha^2} \left(\partial_k U(\alpha, \delta) \right) \big|_{\alpha=\alpha_0, \delta=\delta_0} \\
\partial_k \bar{\lambda}_d &= \frac{d^2}{d\delta^2} \left(\partial_k U(\alpha, \delta) \right) \big|_{\alpha=\alpha_0, \delta=\delta_0} \quad (5.11) \\
\partial_k \bar{\lambda}_{ad} &= \frac{d^2}{d\alpha d\delta} \left(\partial_k U(\alpha, \delta) \right) \big|_{\alpha=\alpha_0, \delta=\delta_0} .
\end{aligned}
$$

These formulas are also valid if one of the symmetries remains unbroken in which case one has to set either α_0 or δ_0 to zero. As long as a one of the symmetries is unbroken, a nonvanishing mass term is associated to it whose flow equation is given by

$$\partial_k \bar{m}_a^2 = \frac{d}{d\alpha} \left(\partial_k U(\alpha, \delta) \right) \big|_{\alpha=0, \delta=\delta_0} \quad (5.12)$$

or

$$\partial_k \bar{m}_d^2 = \frac{d}{d\delta} \left(\partial_k U(\alpha, \delta) \right) \big|_{\alpha=\alpha_0, \delta=0}, \quad (5.13)$$

depending on which of the mass terms remains nonzero.

While in the symmetric regime the fermionic contributions to the flow of the mass terms are always negative and drive the masses towards zero, in the regimes with spontaneous symmetry breaking they may change sign. In this case even the fermionic contribution can lead to an *increase* of the mass terms of the bosonic fields with vanishing order parameters. In particular, if the antiferromagnetic order parameter acquires a nonzero value, this may change the sign of the fermionic contribution to the flow of the superconducting mass term $\bar{m}_d^2$ and prevent it from becoming zero. This effect is shown in Fig. 5.1. In that sense the presence of antiferromagnetic order order in the system has a tendency to prevent the establishment of d-wave superconducting order. Similarly, in the regimes where both α_0 and δ_0 are nonzero, a large value of α_0 has a diminishing influence on the fermionic contribution to the flow of δ_0, which therefore grows less quickly or decreases faster for $k \to 0$ than if α_0 were zero. This effect acts against the coexistence of antiferromagnetic and d-wave superconducting order. Indeed, the phase diagram in Fig. 5.5 shows no region of coexistence of both orders, in contrast to what one might have expected from the flow of the masses and Yukawa couplings in the symmetric regime and the pseudocritical temperatures.

To derive the flow equations for the field expectation values α_0 and δ_0 one can use the condition that $U(\alpha_0, \delta_0)$ is a minimum of the effective potential $U(\alpha, \delta)$. From the fact that the necessary condition

$$\partial_\alpha U(\alpha_0, \delta_0) = \partial_\delta U(\alpha_0, \delta_0) = 0, \quad (5.14)$$

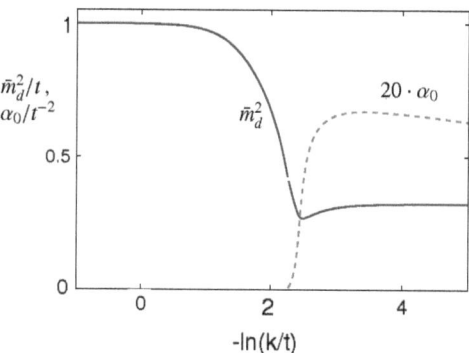

Figure 5.1: Flow of the d-wave superconducting mass term $\bar{m}_d^2$ (solid) and (unrenormalized) antiferromagnetic order parameter α_0 (dashed), the latter multiplied by twenty. Here nonzero α_0 inverts the sign of the fermionic contribution to the flow of $\bar{m}_d^2$ so that it no longer decreases but rather increases and later saturates during the flow. Parameters chosen are $U/t = 3$, $t'/t = -0.1$, $\mu/t = -0.65$ and $T/t = 0.04$

has to hold on all scales k of the renormalization flow, one obtains the prescription

$$\frac{d}{dk}\partial_\alpha U(\alpha_0, \delta_0) = \frac{d}{dk}\partial_\delta U(\alpha_0, \delta_0) = 0 \,. \quad (5.15)$$

Together with Eq. (5.11) the flow equations for the order parameters follow:

$$\begin{aligned}
\partial_k \alpha_0 &= -\frac{\bar{\lambda}_d}{\bar{\lambda}_a \bar{\lambda}_d - \bar{\lambda}_{ad}^2}\partial_\alpha \partial_k U_{B,k}(\alpha, \delta)\big|_{\alpha=\alpha_0,\delta=\delta_0} \\
&\quad + \frac{\bar{\lambda}_{ad}}{\bar{\lambda}_a \bar{\lambda}_d - \bar{\lambda}_{ad}^2}\partial_\delta \partial_k U_{B,k}(\alpha, \delta)\big|_{\alpha=\alpha_0,\delta=\delta_0}, \\
\partial_k \delta_0 &= -\frac{\bar{\lambda}_a}{\bar{\lambda}_a \bar{\lambda}_d - \bar{\lambda}_{ad}^2}\partial_\delta \partial_k U_{B,k}(\alpha, \delta)\big|_{\alpha=\alpha_0,\delta=\delta_0} \\
&\quad + \frac{\bar{\lambda}_{ad}}{\bar{\lambda}_a \bar{\lambda}_d - \bar{\lambda}_{ad}^2}\partial_\alpha \partial_k U_{B,k}(\alpha, \delta)\big|_{\alpha=\alpha_0,\delta=\delta_0}. \quad (5.16)
\end{aligned}$$

For parameter regions where $\bar{\lambda}_a\bar{\lambda}_d - \bar{\lambda}_{ad}^2$ reaches zero the polynomial approximation for the flowing potential $U_B(\alpha, \delta)$ is no longer appropriate, an issue which is discussed below.

For the studied temperature regime $T > T_{min}$ the lowest Matsubara mode dominates in the spontaneously broken regimes ($k < k_{SSB}$) and the

5.2 Flow of the Effective Potential

dimensionality of the problem is effectively reduced from $2+1$ to 2, a mechanism which is known as "dimensional reduction". For computational simplicity, contributions from all bosonic Matsubara modes except the lowest ones are therefore neglected in the spontaneously broken regimes. This assumption becomes exact in the limit of low k, and it is expected to involve only a small quantitative inaccuracy at scales close to the critical scale. If α_0 or δ_0, but not both, are nonzero, the long-range behavior of the system at finite temperature can be described by the $O(3)$-symmetric [74] or $O(2)$-symmetric linear σ-model, depending on whether α_0 or δ_0 is nonzero. The properties of these models are well-known and well understood.

In a regime where the lowest Matsubara mode dominates over the others by far, one is dealing with an effectively two-dimensional problem. In accordance with the Mermin Wagner theorem, the unrenormalized field expectation values α_0 and δ_0 thus have to vanish in the infrared limit $k \to 0$. For the two-dimensional $O(2)$-symmetric model, however, which can be used as an approximation in the dimensionally reduced regime for superconducting order, the *renormalized* field expectation value $\kappa_d = t^2 A_d \delta_0 / T$ may remain nonzero even if δ_0 drops to zero as the gradient coefficient A_d may diverge in this case [75]. This behavior is characteristic of a Kosterlitz-Thouless phase transition [76], for functional renormalization group treatments see [33, 77, 78]. Although the polynomial expansion of the effective potential in Eq. (5.1) is not sufficient to account for the finiteness of κ_d down to $k=0$, it is sufficiently accurate to describe its being nonzero down to scales $k \ll k_{ph}$ much smaller than any realistic inverse probe size l^{-1}.

In terms of the renormalized quantities introduced above, when the fermions are fully gapped and dimensional reduction is efficient so that $k \ll 2\pi T$ and $k \ll \pi$, the flow equations for the order parameters and quartic couplings at vanishing $\bar{\lambda}_{ad} = 0$ reduce to those familiar from the $O(2)$- and $O(3)$-symmetric linear σ-models, namely

$$k\partial_k \kappa_a = \frac{(4-\eta_a)}{16\pi}\left(\frac{3}{(1+2\lambda_a\kappa_a)^2}+2\right) - \eta_a \kappa_a, \quad (5.17)$$

$$k\partial_k \lambda_a = \lambda_a^2 \frac{(4-\eta_a)}{8\pi}\left(\frac{9}{(1+2\lambda_a\kappa_a)^3}+2\right) - 2(1-\eta_a)\lambda_a \quad (5.18)$$

for the a-boson, and

$$k\partial_k \kappa_d = \frac{(4-\eta_d)}{16\pi}\left(\frac{3}{(1+2\lambda_d\kappa_d)^2}+1\right) - \eta_d \kappa_d, \quad (5.19)$$

$$k\partial_k \lambda_d = \lambda_d^2 \frac{(4-\eta_d)}{8\pi}\left(\frac{9}{(1+2\lambda_d\kappa_d)^3}+1\right) - 2(1-\eta_d)\lambda_d \quad (5.20)$$

for the d-boson. Since in the regime with two nonzero order parameters the absolute value of λ_{ad} is normally driven to zero much faster than the two other quartic couplings λ_a and λ_d, the flow of κ_a, κ_d, λ_a, λ_d is generally well described by Eqs. (5.17)-(5.20). If, however, $|\lambda_{ad}|$ is larger than the geometric mean of λ_a and λ_d, i. e. if $|\lambda_{ad}| > \sqrt{\lambda_a \cdot \lambda_d}$, the effective potential $U(\alpha, \delta)$ no longer has a minimum at (α_0, δ_0) and this signals the breakdown of our truncation which relies on an expansion of $U(\alpha, \delta)$ around (α_0, δ_0), assumed to be the location of a minimum. Fortunately, however, the numerical results obtained for the truncation Eq. (5.1) yield a violation of the condition $|\lambda_{ad}| < \sqrt{\lambda_a \cdot \lambda_d}$ only in regions where antiferromagnetism strongly dominates over d-wave superconductivity. In this regime, the effect of d-wave superconducting fluctuations on the emergence of antiferromagnetic order is negligible and the truncation Eq. (5.1) is not natural. Consequently, if in this regime $|\lambda_{ad}|$ rises above $\sqrt{\lambda_a \cdot \lambda_d}$, λ_{ad} is set to zero on all scales whereby the expansion for the effective potential becomes again well-defined.

The main difference between the flow equations for κ_a and λ_a on the one hand and κ_d and λ_d on the other concerns the "+2" in Eqs. (5.17) and (5.18) as opposed to the "+1" in Eqs. (5.19) and (5.20). This corresponds to the different numbers 2 and 1 of Goldstone bosons in the symmetry broken phases of the $O(3)$- and $O(2)$-symmetric linear σ-models, respectively. Since in the presence of a non-negligible order parameter the Goldstone modes have a much stronger influence than the radial modes in driving the order parameter to zero, their number is crucial for how long (in terms of the renormalization group flow) the system remains in the symmetry broken regime.

In order to obtain the anomalous dimensions, one has to determine the flow equations for A_a and A_d in the presence of nonzero κ_a and/or κ_d. To this end, a second derivative of the loop contributions to $P_a(Q)$ and $P_d(Q)$ with respect to spatial momentum is taken and a derivative with respect to the scale k:

$$\partial_k A_a = \partial_k \left(\lim_{l \to 0} \frac{1}{2} \frac{\partial^2}{\partial l^2} \Delta P_a(0, l, 0) \right), \tag{5.21}$$

$$\partial_k A_d = \partial_k \left(\lim_{l \to 0} \frac{1}{2} \frac{\partial^2}{\partial l^2} \Delta P_d(0, l, 0) \right). \tag{5.22}$$

In the regimes exhibiting spontaneous symmetry breaking, the fermionic contributions to η_a and η_d quickly become negligible as soon as the scale drops below the temperature, and it suffices to consider the bosonic contributions. In case the two bosons can independently be described by the two-dimensional $O(3)$- and $O(2)$-symmetric models these contributions are, assuming dimensional reduction,

$$\eta_{a,d} = \frac{1}{\pi} \frac{\lambda_{a,d}^2 \kappa_{a,d}}{(1 + 2\lambda_{a,d} \kappa_{a,d})^2}. \tag{5.23}$$

5.3 Numerical results

In the presence of nonzero λ_{ad}, this formula has to be generalized, yielding

$$\eta_a = \frac{1}{\pi}\left(\lambda_{ad}^2 \frac{\kappa_d(1 - 4\kappa_a(\lambda_a - \kappa_d\lambda_{ad}^2 + 2\kappa_d\lambda_a\lambda_d))}{\left(1 + 2\kappa_d\lambda_d + 2\kappa_a(\lambda_a - 2\kappa_d\lambda_{ad}^2 + 2\kappa_d\lambda_a\lambda_d)\right)^2}\right.$$
$$\left. + \frac{\kappa_a\lambda_a^2(1 + 2\kappa_d\lambda_d)^2}{\left(1 + 2\kappa_d\lambda_d + 2\kappa_a(\lambda_a - 2\kappa_d\lambda_{ad}^2 + 2\kappa_d\lambda_a\lambda_d)\right)^2}\right), \quad (5.24)$$

$$\eta_d = \frac{1}{\pi}\left(\lambda_{ad}^2 \frac{\kappa_a(1 - 4\kappa_d(\lambda_d - \kappa_a\lambda_{ad}^2 + 2\kappa_a\lambda_a\lambda_d))}{\left(1 + 2\kappa_d\lambda_d + 2\kappa_a(\lambda_a - 2\kappa_d\lambda_{ad}^2 + 2\kappa_d\lambda_a\lambda_d)\right)^2}\right.$$
$$\left. + \frac{\kappa_d\lambda_d^2(1 + 2\kappa_a\lambda_a)^2}{\left(1 + 2\kappa_d\lambda_d + 2\kappa_a(\lambda_a - 2\kappa_d\lambda_{ad}^2 + 2\kappa_d\lambda_a\lambda_d)\right)^2}\right), \quad (5.25)$$

which reduces to Eq. (5.23) for $\lambda_{ad} = 0$.

5.3 Numerical results

I now turn to the discussion of the numerical results for the order parameters and the gaps as functions of temperature (Fig. 5.2), the flow of the running couplings in a selected number of cases (Figs. 5.3 and 5.4), and the more general features of the phase diagram (Fig. 5.5) in the symmetry broken regimes SSBa, SSBd and SSBad.

5.3.1 Order parameters

Fig. 5.2 displays the renormalized antiferromagnetic and d-wave superconducting order parameters κ_a and κ_d as functions of temperature at different values of the chemical potential μ for $U = 3t$ and $t' = -0.1t$. Both κ_a and κ_d are evaluated at $k_{ph} = 10^{-9}t \approx 1\,\text{cm}^{-1}$, corresponding to a realistic inverse probe size. The upper left panel shows the temperature dependence of $\hat{\kappa}_a = \kappa_a|_{k=k_{ph}}$ at the van Hove filling $\mu = 4t'$. The shape of this curve for $\hat{\kappa}_a$ is similar to that of the curve presented Fig. 1 in [31] for $t' = \mu = 0$. The temperatures where $\hat{\kappa}_a$ is nonzero, however, are lower according to the results presented here since more fluctuations have been included which have a tendency to destroy antiferromagnetic order. The upper right panel of Fig. 5.2 shows the temperature dependence of $\hat{\kappa}_d = \kappa_d|_{k=k_{ph}}$ at $\mu/t = -0.81$, where only d-wave superconducting order and no antiferromagnetic order occurs. The steep fall to zero of $\hat{\kappa}_d$ at $T = T_c$ can be seen as a remnant of the jump in the superfluid density found for a Kosterlitz-Thouless phase transition at T_c in an improved truncation [78, 33].

The lower panel of Fig. 5.2 shows a situation where, at $\mu/t = -0.72$, one finds nonzero κ_d at low temperatures and nonzero κ_a at higher temperatures. In between, a small temperature region around $T = 0.0125t$ is observed

Functional Renormalization for the Spontaneously Broken Regimes

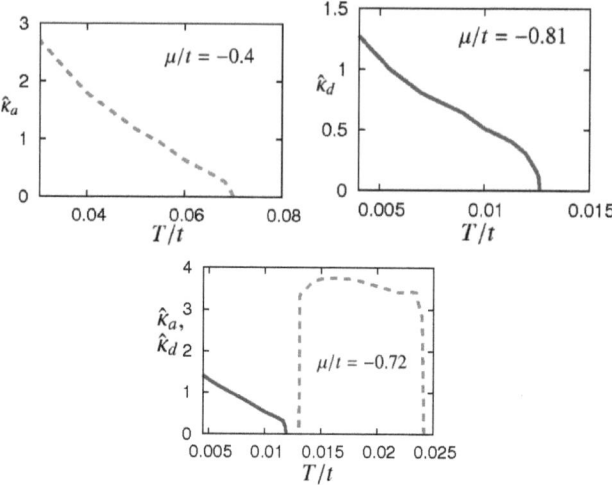

Figure 5.2: Renormalized order parameters $\hat{\kappa}_a$ and $\hat{\kappa}_d$ at the "macroscopic" scale $k = k_{ph}$, corresponding to an inverse probe size of ≈ 1 cm, as a function of temperature T for $U/t = 3$ and different values of μ. The upper left panel shows the temperature dependence $\hat{\kappa}_a$ for $\mu/t = -0.4$, the upper right panel shows $\hat{\kappa}_d$ for $\mu/t = -0.81$. The lower panel shows $\hat{\kappa}_d$ and $\hat{\kappa}_a$ for $\mu/t = -0.72$, where they are nonzero in different temperature ranges.

where neither of the two order parameters remains nonzero down to $k = k_{ph}$. In this region only local but no long-range order is present. An interesting feature of this figure is the steepness of the rise of $\hat{\kappa}_a$ at $T = T_c$, which contrasts with the behavior at van Hove filling (upper panel of Fig. 5.2) where the rise below T_c is relatively smooth. The main reason for this feature is the strong growth with temperature of the final value of λ_a in SYM close to $T = T_c$. This value has an important influence on the initial growth of κ_a in the spontaneously broken regime and therefore at its value at $k = k_{ph}$. This effect is mainly responsible for the smallness of the temperature interval in which $\hat{\kappa}_a$ drops to zero as the temperature approaches T_c from below.

The fact that both order parameters become zero during the flow at values of $k > k_{ph}$ in a temperature region around $T = 0.0125t$ results from the mutual negative influence of the two types of order on each other. This influence is further illustrated by the left and right panels of Fig. 5.3, where the left panel shows the flow of κ_a together with that of κ_d down to $k = k_{ph}$. For the temperature $T = 0.0131t$ used in this graph κ_a is nonzero during a

5.3 Numerical results

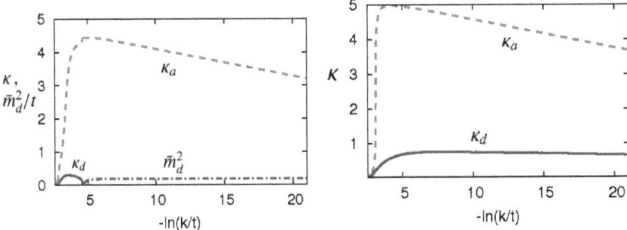

Figure 5.3: Left panel: Flow of the renormalized antiferromagnetic order parameter κ_a (dashed) and the renormalized d-wave superconducting order parameter κ_d (solid) at $U/t = 3$, $t'/t = -0.1$, $\mu/t = -0.72$ and $T/t = 0.0131$. The dashed-dotted curve shows the (unrenormalized) d-wave superconducting mass term $\bar{m}_d^2$ when it has become nonzero again. Right panel: Same as left panel, but neglecting the mutual influence of the order parameters, i. e. each order parameter is set to zero in all contributions to the other boson as well as the inter-boson coupling λ_{ad}.

much longer period of the renormalization flow than κ_d, which becomes zero at $-\ln(k/t) \approx 4.6$ so that the superconducting mass term $\bar{m}_d^2$ (dashed-dotted curve) becomes nonzero again. The right panel of Fig. 5.3, in contrast, shows the flow of the same couplings, but in this case the mutual influence of the order parameters has been neglected. This means that each order parameter is set to zero in all contributions to the other boson and the inter-bosonic quartic coupling λ_{ad} is set to zero. As described in the previous section, this is equivalent to deriving the bosonic contributions to the flow equations from the $O(3)$- and $O(2)$-symmetric linear σ-models at finite temperature. According to this simplified treatment, neglecting the mutual influence of the two types of order, both order parameters remain nonzero down to $k = k_{ph}$. Such a result would suggest a region of coexistence of "global" antiferromagnetic and d-wave superconducting order.

In the present example this coexistence is destroyed by the mutual influence of the antiferromagnetic and superconducing bosons. It is precisely this type of influence which has been taken into account in the left panel of Fig. 5.3. One can therefore conclude that the two types of order have a tendency to destroy each other. A renormalized mean field treatment (See [64], in particular Figs. 10 and 11) seems to suggest an analogous tendency of antiferromagnetism and superconductivity to mutually suppress each other.

For the curves shown in Fig. 5.4 the temperature has been reduced in comparison to Fig. 5.3, so that both order parameters (left panel) are nonzero for an important interval of the flow. Now d-wave superconducting order persists down to much lower scales k of the renormalization flow. Although at an intermediate stage of the flow κ_a is considerably larger than

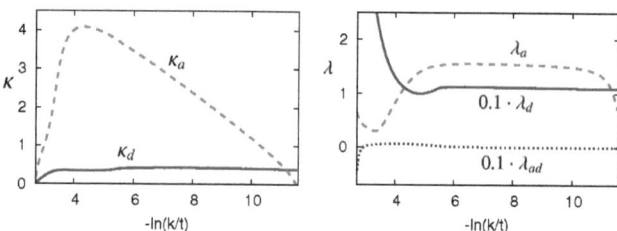

Figure 5.4: Left panel: Flow of the renormalized antiferromagnetic order parameter κ_a (dashed) and renormalized d-wave superconducting order parameter κ_d (solid) at $U/t = 3$, $t'/t = -0.1$, $\mu/t = -0.72$ and $T/t = 0.0118$. Right panel: Flow of the quartic couplings λ_a (dashed), λ_d (solid) and λ_{ad} (dotted), the latter two multiplied by a factor of 0.1, for the same choice of parameters.

κ_d, it vanishes earlier during the flow due to the larger number of Goldstone modes for antiferromagnetism. In the right panel of Fig. 5.4 the flow of the quartic couplings λ_a, λ_d and λ_{ad} is displayed, where λ_{ad} approaches zero much more quickly than λ_a and λ_d so that the two bosons are more or less independent and the flow is dominated by their Goldstone modes at low scales.

Taking things together, the example shown in the lower panel of Fig. 5.2 demonstrates that the phase transitions cannot always be understood by the universal behavior of linear or non-linear uncoupled σ-models. For example, in the $O(3)$-σ-model it is not possible to find a restoration of disorder at temperatures below the ones for which long-range order is realized. The competition of different bosons is crucial for a quantitative understanding of the phase diagram.

5.3.2 Phase diagram

I now turn to the discussion of the phase diagram obtained for $U = 3t$ and $t' = -0.1t$, as shown in Fig. 5.5. For values of $-t'$ which are substantially smaller than $0.1t$ the quartic coupling λ_a eventually becomes negative during the flow, which may indicate a tendency towards a first order transition which is not captured in the present truncation of the effective potential (5.1). For values of $-t'$ which are considerably larger, in contrast, the system exhibits a tendency towards ferromagnetism [28]. In order to account for this instability, the truncation for the effective action and the parameterization of the bosonic propagators and Yukawa couplings specified in Chapter 4 would have to be adjusted accordingly. Upon small variations of t' the qualitative picture of the phase diagram remains essentially unchanged. If

5.3 Numerical results

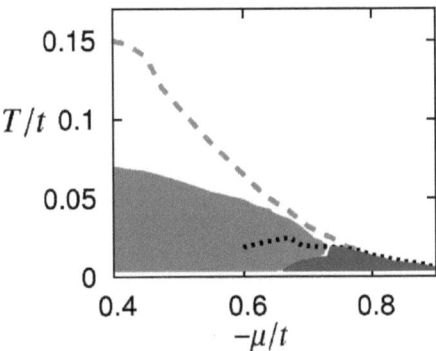

Figure 5.5: Phase diagram for $U/t = 3$ and $t'/t = -0.1$. The solid region for smaller values of $|\mu|$ denotes the antiferromagnetic phase, the solid region for larger values of $|\mu|$ the d-wave superconducting phase. The dashed line and the dotted line indicate the pseudocritical temperatures below which local order sets in for antiferromagnetism and d-wave superconductivity, respectively. The pseudocritical line for antiferromagnetism ends at $-\mu/t \approx 0.79$. The region below $T_{min} = 4 \cdot 10^{-3} t$ has been left blank since calculations were only done for higher temperatures.

$-t'$ is reduced, all phase boundaries are shifted towards smaller values of $-\mu$, if $-t'$ is increased, they move into the other direction. For smaller values of the Hubbard interaction U, critical temperatures are lower and the phase boundaries are shifted in the direction of the van Hove filling chemical potential $\mu = 4t'$. The results obtained for calculations at values of U and t' other than $U = 3t$ and $t' = -0.1t$ do not alter the picture described in what follows.

At the van Hove filling one finds a sizeable difference between the pseudocritical temperature T_{pc} and the true critical temperature T_c for antiferromagnetism, which differ by a factor of about 2, mainly due to the two antiferromagnetic Goldstone modes. In the d-wave superconducting regime at $-\mu/t > 0.75$, in contrast, there is only a slight difference between T_{pc} and T_c, in accordance with earlier results on the $O(2)$-symmetric model [33]. The non-negligible difference between T_{pc} and T_c for d-wave superconductivity in the region between $-\mu/t = 0.66$ and $-\mu/t = 0.75$ is not due to Goldstone fluctuations but arises from the influence of antiferromagnetic order on the flow of κ_d.

One of the most intriguing questions about the phase diagram of the two-dimensional Hubbard model is whether it exhibits a region in parameter space where antiferromagnetic and d-wave superconducting order coexist. In principle the setup employed in the present work allows to assess this question, but the results obtained by means of the truncation used here do not permit a definite answer. While they clearly suggest that the two types of order "do not like each other", they can hardly be taken to rule out the existence of a region in parameter space where antiferromagnetism and d-wave superconductivity coexist. Where the two phases border each other at low temperatures around $-\mu/t = 0.66$ in Fig. 5.5 there is always at least one type of order which persists down to $k = k_{ph} = 1\,\text{cm}^{-1} \approx 10^{-9}t$, and the value of k where the second order parameter drops to zero is often very close to k_{ph}. Coexistence might occur, even within the truncation used here at temperatures below $T_{min} = 4 \cdot 10^{-3} t$, the lowest temperature for which I have done calculations. In all cases where both order parameters remain finite for a considerable part of the flow, the values of the running couplings at $k = k_{ph}$ are highly sensitive to their values at the onset of the spontaneously broken regime. Therefore, it is to be expected that further extensions of the truncation, which may influence the flow on intermediate scales, can have an important effect on the shape of the phase boundaries where the antiferromagnetic and d-wave superconducting phases are close to each other. Self-energy corrections, higher order bosonic couplings and the effect of the antiferromagnetic incommensurability, which have been neglected here in the SSB regimes, may be responsible for whether there exists a region in the phase diagram where antiferromagnetic and d-wave superconducting order coexist at $k = k_{ph}$. The results presented here, however, suggest that *if* there is a region in $\mu - T$-space where antiferromagnetism and d-wave superconductivity coexist on a macroscopic level, this region is probably not very extended.

Chapter 6

Summary and Outlook

The two-dimensional Hubbard model on a square lattice is widely accepted as a basic description of the CuO_2-planes in the high-T_c cuprates. In particular, it is hoped by many that the model might throw some light on the mechanism of Cooper pair formation in these materials. Although the Hubbard Hamiltonian is extraordinarily simple, containing besides the kinetic (hopping) term only a local onsite repulsion U, to solve the model exactly is probably impossible and to establish the most important features of its phase diagram is already very difficult. Renormalization group techniques are among the most promising tools for detecting the leading instabilities of the model in different parameter regimes. In the past decade, the so-called functional renormalization group approach has proved to be a very flexible way of spelling out this idea. In the present work the functional renormalization group scheme has been used in combination with the concept of partial bosonization.

In order to avoid the so-called mean field ambiguity, which arises from the fact that a four-fermion interaction term can be decomposed into different bosonic channels, the concept of partial bosonization is set up in the present work in the form of *flowing bosonization*, a scale-dependent variation of the concept of a Hubbard-Stratonovich transformation which distributes contributions to the four-fermion vertex onto the Yukawa couplings between fermions on bosons depending on their momentum and spin dependence. Taking into account bosonic fields which correspond to the magnetic and charge density as well as s- and d-wave superconducting channels, this procedure allows for an efficient parameterization of the four-fermion vertex that is exact up to one-loop perturbative order. The same setting can also be used to describe the generation of a coupling in the d-wave superconducting channel, technically arising from the scale derivative of the particle-particle box diagram with internal antiferromagnetic propagator lines. The "direct" contribution to the flow of the d-wave Yukawa coupling which has an internal antiferromagnetic line is characterized by a similar momentum structure as the right hand side of the BCS-gap equation. The enhancement of the d-wave Yukawa coupling resulting from this contribution confirms the scenario of d-wave superconductivity as arising from antiferromagnetic spin fluctuations, which successfully predicts the d-wave symmetry of the superconducting order parameter for the cuprates.

Partial bosonization not only allows for an efficient parameterization of the fermionic four-point vertex, but it also enables one to follow the renormalization flow of the running couplings into the regimes where one or more symmetries of the Hamiltonian are spontaneously broken. Entering these regimes is a necessary prerequisite if one wants to compute directly the critical temperatures where (e.g. magnetic or superconducting) order emerges at length scales corresponding to realistic sample sizes. The existence of global order is chiefly determined by fluctuations stemming from the Goldstone modes whose number depends on the order parameter in question. While for antiferromagnetism this number is two, for superconductivity it is only one, resulting in a larger difference between the temperature where local order sets in and the temperature where order is present at macroscopic scales in the case of antiferromagnetism than in the case of superconductivity.

Although the method employed in the present work leads to qualitatively and arguably also semiquantitatively plausible results for the phase diagram of the two-dimensional Hubbard model, there is room for improving the approximations made. In order to obtain a more rigorous treatment of the fermionic-four point-vertex the virtues of the partially bosonized approach used here may be combined with those of the N-patch scheme in the efficient parameterization method developed in [44]. (For an alternative approach using the language of Nambu vertices see [79].) In other regions of the parameter space, for instance at larger values of $-t'$, other bosonic fields would have to be taken into account. These may include a (triplet) p-wave superconducting boson field or a d-wave charge density boson field, corresponding to the so-called Pomeranchuk instability. In a next step, self-energy corrections could be taken into account in a more extensive and systematic manner. For recent work in this direction, based on the efficient parameterization method of [44], see [70].

The truncation for the symmetry broken phases might also fruitfully be extended. Besides taking into account an additional number of bosons, it might be rewarding to go beyond the polynomial approximation for the effective potential made in this work, for instance by discretizing the space of field expectation values and solving the flow equation for the effective potential on a lattice. This would enable an investigation of possible first order phase transitions of the system and of the influence of bosonic fluctuations on these, where the expectation would be that first order phase transition may be turned into second order transitions by bosonic fluctuations [80]. As a starting point, one might consider the bosonic $O(N) \times O(M)$-linear

σ-models in which either one or two symmetries can be spontaneously broken so that more than one order parameter can acquire a nonzero value. Finally, one might try to account for the effect of the antiferromagnetic incommensurability on the flow of the order parameters in the differently ordered regimes.

Appendix A

Notational Conventions

In this work natural units are adopted, i.e. $\hbar = c = k_B = 1$.

A compact notation is employed in which for $n \in \mathbb{Z}$ one has $Q = (\omega_n = 2\pi n T, \mathbf{q})$ for bosonic and $Q = (\omega_n = (2n+1)\pi T, \mathbf{q})$ for fermionic fields and

$$\sum_X = \int_0^\beta d\tau \sum_{\mathbf{x}}, \qquad \sum_Q = T \sum_{n=-\infty}^{\infty} \int_{-\pi}^{\pi} \frac{d^2q}{(2\pi)^2}, \qquad (A.1)$$

$$\delta(X - X') = \delta(\tau - \tau')\delta_{\mathbf{x},\mathbf{x}'}, \qquad \delta(Q - Q') = T^{-1}\delta_{n,n'}(2\pi)^2\delta^{(2)}(\mathbf{q} - \mathbf{q}').$$

The components of the momentum $\mathbf{q}$ are measured in units of the inverse lattice distance a^{-1}, which is set to one. The discreteness of the lattice is reflected by the 2π-periodicity of the momenta $\mathbf{q}$.

Fourier transforms are defined through the convention

$$\hat{\psi}(X) = \sum_Q \hat{\psi}(Q)e^{iQX}, \qquad \hat{\psi}(X)^\dagger = \sum_Q \hat{\psi}^\dagger(Q)e^{-iQX}, \qquad (A.2)$$

for fermionic, and analogously for bosonic, fields.

Vectors in frequency-momentum space that are frequently used in the evaluation of vertex functions are given by

$$\begin{aligned} 0 &= (0,0,0), & 0^+ &= (2\pi T, 0, 0), \\ \Pi &= (0, \pi, \pi), & \Pi^+ &= (2\pi T, \pi, \pi), \\ L &= (\pi T, \pi, 0), & L' &= (\pi T, 0, \pi). \end{aligned} \qquad (A.3)$$

Other frequently used vectors involve the antiferromagnetic incommensurability $\hat{q}$,

$$\begin{aligned} \hat{Q}_x &= (0, \mathbf{q}_1) = -(0, \mathbf{q}_2) = (0, \hat{q}, 0) \\ \hat{Q}_y &= (0, \mathbf{q}_3) = -(0, \mathbf{q}_4) = (0, 0, \hat{q}). \end{aligned} \qquad (A.4)$$

Appendix B

Pauli Matrices and Spin Projections

In order to compute the flow equations for the individual running couplings from the flow equation for the flowing action, relations between several tensors and matrices are needed. Some are given in this appendix to facilitate the understanding of the calculations which lead to the flow equations given in the text.

The Pauli matrices, which form a basis of the space of traceless Hermitean 2×2-matrices, are given by

$$\sigma^1 = \begin{pmatrix} 0 & 1 \\ 1 & 0 \end{pmatrix}, \quad \sigma^2 = \begin{pmatrix} 0 & -i \\ i & 0 \end{pmatrix}, \quad \sigma^3 = \begin{pmatrix} 1 & 0 \\ 0 & -1 \end{pmatrix}, \quad (\text{B.1})$$

and the vector of the three Pauli matrices is denoted by $\boldsymbol{\sigma}$.

The Pauli matrices obey $\sigma^i = (\sigma^i)^\dagger = (\sigma^i)^{-1}$ for $i = 1, 2, 3$, and they obey the (anti-) commutation relations

$$\{\sigma^i, \sigma^j\} = 2\delta^{ij}, \quad [\sigma^i, \sigma^j] = 2i\epsilon^{ijk}\sigma^k, \quad (\text{B.2})$$

where ϵ^{ijk} is the three-dimensional Levi-Civita symbol.

The singlet and triplet projectors $S_{\alpha\gamma;\beta\delta}$ and $T_{\alpha\gamma;\beta\delta}$ are defined as

$$S_{\alpha\gamma;\beta\delta} = \frac{1}{2}(\delta_{\alpha\beta}\delta_{\gamma\delta} - \delta_{\alpha\delta}\delta_{\gamma\beta}), \quad (\text{B.3})$$

$$T_{\alpha\gamma;\beta\delta} = \frac{1}{2}(\delta_{\alpha\beta}\delta_{\gamma\delta} + \delta_{\alpha\delta}\delta_{\gamma\beta}). \quad (\text{B.4})$$

$$(\text{B.5})$$

For the computation of one-loop diagrams the following identities are useful:

$$(\sigma^i\sigma^j)_{\alpha\beta}(\sigma^i\sigma^j)_{\gamma\delta} = 9S_{\alpha\gamma;\beta\delta} + T_{\alpha\gamma;\beta\delta}, \quad (\text{B.6})$$

$$\sigma^j_{\alpha\beta}\sigma^j_{\gamma\delta} = -3S_{\alpha\gamma;\beta\delta} + T_{\alpha\gamma;\beta\delta}, \quad (\text{B.7})$$

$$\delta_{\alpha\beta}\delta_{\gamma\delta} = S_{\alpha\gamma;\beta\delta} + T_{\alpha\gamma;\beta\delta}, \quad (\text{B.8})$$

$$(\sigma^i\sigma^j)_{\alpha\delta}(\sigma^j\sigma^j)_{\gamma\beta} = \frac{9}{2}\delta_{\alpha\beta}\delta_{\gamma\delta} + \frac{1}{2}\sigma^j_{\alpha\beta}\sigma^j_{\gamma\delta}, \tag{B.9}$$

$$\sigma^j_{\alpha\delta}\sigma^j_{\gamma\beta} = \frac{3}{2}\delta_{\alpha\beta}\delta_{\gamma\delta} - \frac{1}{2}\sigma^j_{\alpha\beta}\sigma^j_{\gamma\delta}, \tag{B.10}$$

$$\delta_{\alpha\delta}\delta_{\gamma\beta} = \frac{1}{2}\delta_{\alpha\beta}\delta_{\gamma\delta} + \frac{1}{2}\sigma^j_{\alpha\beta}\sigma^j_{\gamma\delta}. \tag{B.11}$$

Appendix C

Box diagrams

C.1 Particle-particle Diagrams

The contributions to $\partial_k \bar{h}_s^2$ and $\partial_k \bar{h}_d^2$ that arise from box diagrams having two real boson internal lines (see the first diagram in the last line of Fig. 4.1) are given by:

$$\Delta \Gamma_{F,pp} = -\frac{1}{2} \sum_{K_1,K_2,K_3,K_4} \sum_P \quad \text{(C.1)}$$
$$\left(\frac{\bar{h}_a^2(P)\bar{h}_a^2(K_1 - K_2 + P)(\sigma^i \sigma^j)_{\alpha\beta}(\sigma^i \sigma^j)_{\gamma\delta}}{P_F(K_1 - P - \Pi)\tilde{P}_a(P)\tilde{P}_a(K_1 - K_2 + P)P_F(K_3 + P - \Pi)} \right.$$
$$+ \frac{2\bar{h}_a^2(P)\bar{h}_\rho^2(K_1 - K_2 + P)\sigma^j_{\alpha\beta}\sigma^j_{\gamma\delta}}{P_F(K_1 - P - \Pi)\tilde{P}_a(P)\tilde{P}_\rho(K_1 - K_2 + P)P_F(K_3 + P - \Pi)}$$
$$\left. + \frac{\bar{h}_\rho^2(P)\bar{h}_\rho^2(K_1 - K_2 + P)\delta_{\alpha\beta}\delta_{\gamma\delta}}{P_F(K_1 - P - \Pi)\tilde{P}_\rho(P)\tilde{P}_\rho(K_1 - K_2 + P)P_F(K_3 + P - \Pi)} \right)$$
$$\times \delta(K_1 - K_2 + K_3 - K_4)\, \psi_\alpha^\dagger(K_1)\psi_\beta(K_2)\psi_\gamma^\dagger(K_3)\psi_\delta(K_4).$$

The triplet part is neglected, and the projection onto the singlet part is taken using the identities (B.6), (B.7) and (B.8).

Focusing on the first line of Eq. (C.1) (the others are treated analogously), the resulting loop-correction to the four-point vertex reads (for the singlet part)

$$\Delta \Gamma_{F,pp,s}^{(4),aa}(K_1, K_2, K_3, K_4) = -\frac{9}{4}\sum_P \left(\left(\frac{\bar{h}_a^2(P - K_2)}{\tilde{P}_a(P - K_2)} + \frac{\bar{h}_a^2(P - K_4)}{\tilde{P}_a(P - K_4)} \right) \right.$$
$$\left. \times \left(\frac{\bar{h}_a^2(P + K_1)}{P_F(P - \Pi)P_F(-P + K_1 + K_3 - \Pi)\tilde{P}_a(P + K_1)} \right) \right) \quad \text{(C.2)}$$

At this point, an approximation is required for how this contribution can be distributed onto the s- and d-wave superconducting channels. First it is rewritten by twice adding and subtracting the same term,

$$\begin{aligned}
\Delta\Gamma_{F,pp}^{(4),aa}(K_1,K_2,K_3,K_4) &= -\frac{9}{4}\sum_P \left(\frac{\bar{h}_a^2(P-K_1)}{\tilde{P}_a(P-K_1)} + \frac{\bar{h}_a^2(P+K_1)}{\tilde{P}_a(P+K_1)}\right) \\
&\times \left(\frac{\bar{h}_a^2(P+K_1)}{P_F(P-\Pi)P_F(-P+K_1+K_3-\Pi)\tilde{P}_a(P+K_1)}\right) \\
&+ \left(\left(\frac{\bar{h}_a^2(P-K_2)}{\tilde{P}_a(P-K_2)} - \frac{\bar{h}_a^2(P-K_1)}{\tilde{P}_a(P-K_1)}\right)\right. \\
&\times \left.\left(\frac{\bar{h}_a^2(P+K_1)}{P_F(P-\Pi)P_F(-P+K_1+K_3-\Pi)\tilde{P}_a(P+K_1)}\right)\right) \\
&+ \left(\left(\frac{\bar{h}_a^2(P-K_4)}{\tilde{P}_a(P-K_4)} - \frac{\bar{h}_a^2(P+K_1)}{\tilde{P}_a(P+K_1)}\right)\right. \\
&\times \left.\left(\frac{\bar{h}_a^2(P+K_1)}{P_F(P-\Pi)P_F(-P+K_1+K_3-\Pi)\tilde{P}_a(P+K_1)}\right)\right).
\end{aligned}$$
(C.3)

If $K_1 = K_2$ and $K_1 = -K_4$ or $K_1 = -K_2$ and $K_1 = K_4$, only the first two lines are nonzero. The contribution from the other lines is expected to be most important, in contrast, if the spatial part of K_1 is orthogonal to that of K_2 and K_4, particularly for K_1, K_2, K_3, K_4 taken from the vectors l and l'. A natural approximation, for instance for the third line of (C.3), which allows to bring into play the d-wave form factor f_d is

$$\begin{aligned}
&\left(\frac{\bar{h}_a^2(P-K_2)}{\tilde{P}_a(P-K_2)} - \frac{\bar{h}_a^2(P-K_1)}{\tilde{P}_a(P-K_1)}\right) \\
&\approx \frac{1}{2}\left(f_d^2(K_1) - f_d(K_1)f_d(K_2)\right)\left(\frac{\bar{h}_a^2(P-D_{\pi/2}(K_1))}{\tilde{P}_a(P-D_{\pi/2}(K_1))} - \frac{\bar{h}_a(P-K_1)}{\tilde{P}_a(P-K_1)}\right),
\end{aligned}$$
(C.4)

where $D_\varphi(Q)$ denotes the vector Q after rotation of its spatial momentum part by the angle φ. The fifth line in (C.3) is treated analogously.
For $K_1 = K_2$ and for $K_1 \perp K_2$ (where the sign $\perp$ concerns spatial momentum) the approximation (C.4) is exact. As it smoothly interpolates between these two cases, it can be expected to give a good approximation.
Using that for $L = (\pi T, \pi, 0)$ and $L' = (\pi T, 0, \pi) = D_{\pi/2}(L)$, where the diagram is eventually evaluated, one has $f_d^2(L^{(')}) = 1$, the contribution to

C.1 Particle-particle Diagrams

the effective action can be approximated by

$$\Delta\Gamma^{(4),aa}_{F,pp}(K_1, K_2, K_3, K_4) \approx -\frac{9}{8} \sum_{\epsilon=\pm 1} \sum_P \tag{C.5}$$

$$\left(\frac{\bar{h}_a^2(P+K_1)}{P_F(P-\Pi)P_F(-P+K_1+K_3-\Pi)\tilde{P}_a(P+K_1)} \right.$$

$$\times \left(\frac{\bar{h}_a^2(P+\epsilon K_1)}{\tilde{P}_a(P+\epsilon K_1)} + \frac{\bar{h}_a^2(P+\epsilon D_{\pi/2}(K_1))}{\tilde{P}_a(P+\epsilon D_{\pi/2}(K_1))} \right)$$

$$+ \frac{\bar{h}_a^2(P+K_1)\, f_d(\frac{1}{2}(K_1-K_3))\, f_d(\frac{1}{2}(K_2-K_4))}{P_F(P-\Pi)P_F(-P+K_1+K_3-\Pi)\tilde{P}_a(P+K_1)}$$

$$\left. \times \left(\frac{\bar{h}_a^2(P+\epsilon K_1)}{\tilde{P}_a(P+\epsilon K_1)} - \frac{\bar{h}_a^2(P+\epsilon D_{\pi/2}(K_1))}{\tilde{P}_a(P+\epsilon D_{\pi/2}(K_1))} \right) \right).$$

Comparison with Eqs. (4.8) and (4.9) suggests that the part not involving the d-wave form factors can be absorbed by the s-boson and the part involving the d-wave form factors by the d-boson.

Evaluating at external momenta $L^{(\prime)}$ and inserting the scale derivative operator $\tilde{\partial}_k$ one obtains the contributions to $\partial_k \bar{h}_s$ and $\partial_k \bar{h}_d$ from the box diagram with two internal antiferromagnetic lines as:

$$\left(\partial_k \bar{h}_{s,d}(0)^2\right)^{aa} = \bar{m}_{s,d}^2 \frac{9}{16} \sum_{\epsilon=\pm 1} \sum_P \tilde{\partial}_k \tag{C.6}$$

$$\left(\frac{\bar{h}_a^2(P+L)}{P_F^k(P)P_F^k(-P)\tilde{P}_a^k(P+L)} \left(\frac{\bar{h}_a^2(P+\epsilon L)}{\tilde{P}_a^k(P+\epsilon L)} \pm \frac{\bar{h}_a^2(P+\epsilon L')}{\tilde{P}_a^k(P+\epsilon L')} \right) \right),$$

where the $+$-sign pertains to the flow of the s-wave and the $-$-sign to the flow of the d-wave Yukawa coupling.

In complete analogy, one obtains the contributions to $\partial_k \bar{h}_{s,d}(0)^2$ resulting from the box diagram with one antiferromagnetic and one charge density internal line as

$$\left(\partial_k \bar{h}_{s,d}(0)^2\right)^{a\rho} = \bar{m}_{s,d}^2 \frac{3}{8} \sum_{\epsilon=\pm 1} \sum_P \tilde{\partial}_k \tag{C.7}$$

$$\left(\frac{\bar{h}_\rho^2(P+L)}{P_F^k(P)P_F^k(-P)\tilde{P}_\rho^k(P+L)} \left(\frac{\bar{h}_a^2(P+\epsilon L)}{\tilde{P}_a^k(P+\epsilon L)} \pm \frac{\bar{h}_a^2(P+\epsilon L')}{\tilde{P}_a^k(P+\epsilon L')} \right) \right).$$

Similarly, from the box diagram with two internal charge density lines one obtains

$$\left(\partial_k \bar{h}_{s,d}(0)^2\right)^{\rho\rho} = \bar{m}_{s,d}^2 \frac{1}{16} \sum_{\epsilon=\pm 1} \sum_P \tilde{\partial}_k \tag{C.8}$$

$$\left(\frac{\bar{h}_\rho^2(P+L)}{P_F^k(P)P_F^k(-P)\tilde{P}_\rho^k(P+L)} \left(\frac{\bar{h}_\rho^2(P+\epsilon L)}{\tilde{P}_\rho^k(P+\epsilon L)} \pm \frac{\bar{h}_\rho^2(P+\epsilon L')}{\tilde{P}_\rho^k(P+\epsilon L')} \right) \right).$$

C.2 Particle-hole Diagrams

I now come to the particle-hole box diagrams, starting with those having only real internal bosonic lines. Their correction to the effective action is given by

$$
\begin{aligned}
\Delta\Gamma^{a\rho}_{F,ph} &= \frac{1}{2} \sum_{K_1,K_2,K_3,K_4} \sum_P \\
&\quad \left(\frac{\bar{h}_a^2(P)\bar{h}_a^2(K_3-K_2+P)\,(\sigma^i\sigma^j)_{\alpha\delta}(\sigma^j\sigma^i)_{\gamma\beta}}{P_F(K_1-P-\Pi)\tilde{P}_a(P)\tilde{P}_a(K_3-K_2+P)P_F(K_2-P-\Pi)} \right. \\
&\quad +2 \frac{\bar{h}_a^2(P)\bar{h}_\rho^2(K_3-K_2+P)\,\sigma^j_{\alpha\delta}\sigma^j_{\gamma\beta}}{P_F(K_1-P-\Pi)\tilde{P}_a(P)\tilde{P}_\rho(K_3-K_2+P)P_F(K_2-P-\Pi)} \\
&\quad \left. + \frac{\bar{h}_\rho^2(P)\bar{h}_\rho^2(K_3-K_2+P)\,\delta_{\alpha\delta}\delta_{\gamma\beta}}{P_F(K_1-P-\Pi)\tilde{P}_\rho(P)\tilde{P}_\rho(K_3-K_2+P)P_F(K_2-P-\Pi)} \right) \\
&\quad \times \delta(K_1-K_2+K_3-K_4)\,\psi^\dagger_\alpha(K_1)\psi_\beta(K_2)\psi^\dagger_\gamma(K_3)\psi_\delta(K_4).
\end{aligned}
\tag{C.9}
$$

In order to obtain the resulting contributions to the flow equations, one has to derive with respect to the fermionic fields and to evaluate these at external momenta $L^{(\prime)}$. An average is taken over all contribution for which, in the case of $\bar{h}_a^2(0)$ for example, the condition $K_1 - K_2 = \Pi$ is fulfilled. Furthermore the identities (B.9)-(B.11) are employed and the derivative operator $\tilde{\partial}_k$ is inserted. One obtains for the contribution to the flow equation of $\bar{h}_a^2(0)$

$$
\begin{aligned}
\left(\partial_k \bar{h}_a^2(0)\right)^{a\rho}/\bar{m}_a^2 &= -\frac{1}{4}\sum_P \tilde{\partial}_k \frac{\bar{h}_a^2(P)}{P_F^k(P+L)\tilde{P}_a^k(P)P_F^k(P+L')} \\
&\quad \times \left(\frac{\bar{h}_a^2(P)}{\tilde{P}_a^k(P)} + \frac{\bar{h}_a^2(P)}{\tilde{P}_a^k(P+0^+)} + \frac{\bar{h}_a^2(P+\Pi)}{\tilde{P}_a^k(P+\Pi)} + \frac{\bar{h}_a^2(P+\Pi)}{\tilde{P}_a^k(P+\Pi^+)} \right) \\
&\quad + \frac{1}{2}\sum_P \tilde{\partial}_k \frac{\bar{h}_\rho^2(P)}{P_F^k(P+L)\tilde{P}_\rho^k(P)P_F^k(P+L')} \\
&\quad \times \left(\frac{\bar{h}_a^2(P)}{\tilde{P}_a^k(P)} + \frac{\bar{h}_a^2(P)}{\tilde{P}_a^k(P+0^+)} + \frac{\bar{h}_a^2(P+\Pi)}{\tilde{P}_a^k(P+\Pi)} + \frac{\bar{h}_a^2(P+\Pi)}{\tilde{P}_a^k(P+\Pi^+)} \right) \\
&\quad - \frac{1}{4}\sum_P \tilde{\partial}_k \frac{\bar{h}_\rho^2(P)}{P_F^k(P+L)\tilde{P}_\rho^k(P)P_F^k(P+L')} \\
&\quad \times \left(\frac{\bar{h}_\rho^2(P)}{\tilde{P}_\rho^k(P)} + \frac{\bar{h}_\rho^2(P)}{\tilde{P}_\rho^k(P+0^+)} + \frac{\bar{h}_\rho^2(P+\Pi)}{\tilde{P}_\rho^k(P+\Pi)} + \frac{\bar{h}_\rho^2(P+\Pi)}{\tilde{P}_\rho^k(P+\Pi^+)} \right)
\end{aligned}
\tag{C.10}
$$

C.2 Particle-hole Diagrams

and for that of $\bar{h}_\rho^2(\Pi)$

$$
\begin{aligned}
\left(\partial_k \bar{h}_\rho^2(0)\right)^{a\rho}/\bar{m}_\rho^2 &= -\frac{9}{4}\sum_P \tilde{\partial}_k \frac{\bar{h}_a^2(P)}{P_F^k(P+L)\tilde{P}_a^k(P)P_F^k(P+L')} \\
&\quad \times \left(\frac{\bar{h}_a^2(P)}{\tilde{P}_a^k(P)} + \frac{\bar{h}_a^2(P)}{\tilde{P}_a^k(P+0^+)} + \frac{\bar{h}_a^2(P+\Pi)}{\tilde{P}_a^k(P+\Pi)} + \frac{\bar{h}_a^2(P+\Pi)}{\tilde{P}_a^k(P+\Pi^+)}\right) \\
&\quad -\frac{3}{2}\sum_P \tilde{\partial}_k \frac{\bar{h}_\rho^2(P)}{P_F^k(P+L)\tilde{P}_\rho^k(P)P_F^k(P+L')} \\
&\quad \times \left(\frac{\bar{h}_a^2(P)}{\tilde{P}_a^k(P)} + \frac{\bar{h}_a^2(P)}{\tilde{P}_a^k(P+0^+)} + \frac{\bar{h}_a^2(P+\Pi)}{\tilde{P}_a^k(P+\Pi)} + \frac{\bar{h}_a^2(P+\Pi)}{\tilde{P}_a^k(P+\Pi^+)}\right) \\
&\quad -\frac{1}{4}\sum_P \tilde{\partial}_k \frac{\bar{h}_\rho^2(P)}{P_F^k(P+L)\tilde{P}_\rho^k(P)P_F^k(P+L')} \\
&\quad \times \left(\frac{\bar{h}_\rho^2(P)}{\tilde{P}_\rho^k(P)} + \frac{\bar{h}_\rho^2(P)}{\tilde{P}_\rho^k(P+0^+)} + \frac{\bar{h}_\rho^2(P+\Pi)}{\tilde{P}_\rho^k(P+\Pi)} + \frac{\bar{h}_\rho^2(P+\Pi)}{\tilde{P}_\rho^k(P+\Pi^+)}\right).
\end{aligned}
\tag{C.11}
$$

For the definition of the vectors 0^+ and Π^+ see (A.3). The contributions to $\partial_k \bar{h}_a(\Pi)$ and $\partial_k \bar{h}_\rho(0)$ are obtained through replacing L' by L.

Box diagrams with two internal Cooper pair lines are all of the particle-hole type. Taken together, they are given by

$$
\begin{aligned}
\Delta \Gamma_F^{sd} &= 4 \sum_{K_1, K_2, K_3, K_4} \sum_P \left(\delta_{\alpha\beta}\delta_{\gamma\delta} + \sigma_{\alpha\beta}^j \sigma_{\gamma\delta}^j\right) \\
&\quad \left(\frac{\bar{h}_s^2(P)\bar{h}_s^2(K_3 - K_2 + P)}{P_F(P+K_1)\tilde{P}_s(P)P_F(P+K_2)\tilde{P}_s(K_3 - K_2 + P)}\right. \\
&\quad +\frac{2\bar{h}_s^2(K_3 - K_2 + P)\bar{h}_d^2(P)f_d\left(\frac{1}{2}(P+2K_1)\right)f_d\left(\frac{1}{2}(P+2K_2)\right)}{P_F(P+K_1)\tilde{P}_s(K_3 - K_2 + P)P_F(K_1 - K_2 + P)\tilde{P}_d(P)} \\
&\quad +\frac{\bar{h}_d^2(P)\bar{h}_d(K_3 - K_2 + P)f_d\left(\frac{1}{2}(P+2K_1)\right)f_d\left(\frac{1}{2}(P+2K_2)\right)}{P_F(P+K_1)\tilde{P}_d(K_3 - K_2 + P)P_F(P+K_2)\tilde{P}_d(P)} \\
&\quad \left. \times f_d\left(\frac{1}{2}(P+K_2+K_3)\right)f_d\left(\frac{1}{2}(P+K_1+K_4)\right)\right) \\
&\quad \times \delta\left(K_1 - K_2 + K_3 - K_4\right)\psi_\alpha^\dagger(K_1)\psi_\beta(K_2)\psi_\gamma^\dagger(K_3)\psi_\delta(K_4).
\end{aligned}
\tag{C.12}
$$

Using Eq. (B.11), the contributions from this to $\partial_k \bar{h}_a^2(0)$ and $\partial_k \bar{h}_\rho^2(\Pi)$ are

$$
\begin{aligned}
\left(\partial_k \bar{h}_a^2(0)\right)^{sd} / \bar{m}_a^2 &= \left(\partial_k \bar{h}_\rho^2(\Pi)\right)^{sd} / \bar{m}_\rho^2 = \\
&- \sum_P \tilde{\partial}_k \frac{\bar{h}_s^2(P)}{P_F^k(P+L)\tilde{P}_s^k(P)P_F^k(P+L')} \\
&\times \left(\frac{\bar{h}_s^2(P)}{\tilde{P}_s^k(P)} + \frac{\bar{h}_s^2(P)}{\tilde{P}_s^k(P+0^+)} + \frac{\bar{h}_s^2(P+\Pi)}{\tilde{P}_s^k(P+\Pi)} + \frac{\bar{h}_s^2(P+\Pi)}{\tilde{P}_s^k(P+\Pi^+)} \right) \\
&- 2 \sum_P \tilde{\partial}_k \frac{\bar{h}_d^2(P) f_d(P/2+L)) f_d(P/2+L')}{P_F^k(P+L)\tilde{P}_d^k(P)P_F^k(P+L')} \\
&\times \left(\frac{\bar{h}_s^2(P)}{\tilde{P}_s^k(P)} + \frac{\bar{h}_s^2(P)}{\tilde{P}_s^k(P+0^+)} + \frac{\bar{h}_s^2(P+\Pi)}{\tilde{P}_s^k(P+\Pi)} + \frac{\bar{h}_s^2(P+\Pi)}{\tilde{P}_s^k(P+\Pi^+)} \right) \\
&- \sum_P \tilde{\partial}_k \frac{\bar{h}_d^2(P) f_d(P/2+L)) f_d(P/2+L')}{P_F^k(P+L)\tilde{P}_d^k(P)P_F^k(P+L')} \\
&\times \left(\bar{h}_d^2(P) f_d(P/2+L) f_d(P/2+L') \left(\frac{1}{\tilde{P}_d^k(P)} + \frac{1}{\tilde{P}_d^k(P+0^+)} \right) \right. \\
&\left. + \bar{h}_d^2(P+\Pi) f_d^2(P/2+\Pi/2) \left(\frac{1}{\tilde{P}_d^k(P+\Pi)} + \frac{1}{\tilde{P}_d^k(P+\Pi^+)} \right) \right)
\end{aligned}
$$
(C.13)

The loop corrections to the effective action due to box diagrams with both an internal Cooper pair boson line and an internal real boson line are given by

$$
\begin{aligned}
\Delta \Gamma_F^{as/d} &= -2 \sum_{K_1,K_2,K_3,K_4} \sum_P \left(3\delta_{\alpha\beta}\delta_{\gamma\delta} + \sigma_{\alpha\beta}^j \sigma_{\gamma\delta}^j \right) \\
&\left(\frac{\bar{h}_a^2(K_3 - K_2 + P)\bar{h}_s^2(P)}{P_F(P+K_1)\tilde{P}_s(P)P_F(P+K_2)\tilde{P}_a(K_3 - K_2 + P)} \right. \\
&\left. + \frac{\bar{h}_a^2(K_3 - K_2 + P)\bar{h}_d^2(P) f_d(P/2+K_1)f_d(P/2+K_2)}{P_F(P+K_1)\tilde{P}_d(P)P_F(P+K_2)\tilde{P}_a(K_3 - K_2 + P)} \right), \\
\Delta \Gamma_F^{\rho s/d} &= -2 \sum_{K_1,K_2,K_3,K_4} \sum_P \left(\delta_{\alpha\beta}\delta_{\gamma\delta} - \sigma_{\alpha\beta}^j \sigma_{\gamma\delta}^j \right) \\
&\left(\frac{\bar{h}_\rho^2(K_3 - K_2 + P)\bar{h}_s^2(P)}{P_F(P+K_1)\tilde{P}_s(P)P_F(P+K_2)\tilde{P}_\rho(K_3 - K_2 + P)} \right. \\
&\left. + \frac{\bar{h}_\rho^2(K_3 - K_2 + P)\bar{h}_d^2(P) f_d(P/2+K_1)f_d(P/2+K_2)}{P_F(P+K_1)\tilde{P}_d(P)P_F(P+K_2)\tilde{P}_\rho(K_3 - K_2 + P)} \right)
\end{aligned}
$$
(C.14)

(C.15)

C.2 Particle-hole Diagrams

The resulting contributions from this to $\partial_k \bar{h}_a^2(0)$ and $\partial_k \bar{h}_\rho^2(\Pi)$ are given by

$$\left(\partial_k \bar{h}_a^2(0)\right)^{a/\rho,s/d} = \frac{\bar{m}_a^2}{2} \sum_P \tilde{\partial}_k \tag{C.16}$$

$$\left(\frac{\bar{h}_s^2(P)}{P_F^k(P+L)\tilde{P}_s^k(P)P_F^k(P+L')}\right.$$
$$\times \left(\frac{\bar{h}_a^2(P)}{\tilde{P}_a^k(P)} + \frac{\bar{h}_a^2(P)}{\tilde{P}_a^k(P+0^+)} + \frac{\bar{h}_a^2(P+\Pi)}{\tilde{P}_a^k(P+\Pi)} + \frac{\bar{h}_a^2(P+\Pi)}{\tilde{P}_a^k(P+\Pi^+)}\right)$$
$$+ \frac{\bar{h}_d^2(P) f_d(P/2+L)) f_d(P/2+L')}{P_F^k(P+L)\tilde{P}_d^k(P)P_F^k(P+L')}$$
$$\times \left(\frac{\bar{h}_a^2(P)}{\tilde{P}_a^k(P)} + \frac{\bar{h}_a^2(P)}{\tilde{P}_a^k(P+0^+)} + \frac{\bar{h}_a^2(P+\Pi)}{\tilde{P}_a^k(P+\Pi)} + \frac{\bar{h}_a^2(P+\Pi)}{\tilde{P}_a^k(P+\Pi^+)}\right)$$
$$- \frac{\bar{h}_s^2(P)}{P_F^k(P+L)\tilde{P}_s^k(P)P_F^k(P+L')}$$
$$\times \left(\frac{\bar{h}_\rho^2(P)}{\tilde{P}_\rho^k(P)} + \frac{\bar{h}_\rho^2(P)}{\tilde{P}_\rho^k(P+0^+)} + \frac{\bar{h}_\rho^2(P+\Pi)}{\tilde{P}_\rho^k(P+\Pi)} + \frac{\bar{h}_\rho^2(P+\Pi)}{\tilde{P}_\rho^k(P+\Pi^+)}\right)$$
$$- \frac{\bar{h}_d^2(P) f_d(P/2+L)) f_d(P/2+L')}{P_F^k(P+L)\tilde{P}_d^k(P)P_F^k(P+L')}$$
$$\left.\times \left(\frac{\bar{h}_\rho^2(P)}{\tilde{P}_\rho^k(P)} + \frac{\bar{h}_\rho^2(P)}{\tilde{P}_\rho^k(P+0^+)} + \frac{\bar{h}_\rho^2(P+\Pi)}{\tilde{P}_\rho^k(P+\Pi)} + \frac{\bar{h}_\rho^2(P+\Pi)}{\tilde{P}_\rho^k(P+\Pi^+)}\right)\right),$$

$$\left(\partial_k \bar{h}_\rho^2(\Pi)\right)^{a/\rho,s/d} = \frac{\bar{m}_\rho^2}{2} \sum_P \tilde{\partial}_k \tag{C.17}$$

$$\left(3 \frac{\bar{h}_s^2(P)}{P_F^k(P+L)\tilde{P}_s^k(P)P_F^k(P+L')}\right.$$
$$\times \left(\frac{\bar{h}_a^2(P)}{\tilde{P}_a^k(P)} + \frac{\bar{h}_a^2(P)}{\tilde{P}_a^k(P+0^+)} + \frac{\bar{h}_a^2(P+\Pi)}{\tilde{P}_a^k(P+\Pi)} + \frac{\bar{h}_a^2(P+\Pi)}{\tilde{P}_a^k(P+\Pi^+)}\right)$$
$$+3 \frac{\bar{h}_d^2(P) f_d(P/2+L)) f_d(P/2+L')}{P_F^k(P+L)\tilde{P}_d^k(P)P_F^k(P+L')}$$
$$\times \left(\frac{\bar{h}_a^2(P)}{\tilde{P}_a^k(P)} + \frac{\bar{h}_a^2(P)}{\tilde{P}_a^k(P+0^+)} + \frac{\bar{h}_a^2(P+\Pi)}{\tilde{P}_a^k(P+\Pi)} + \frac{\bar{h}_a^2(P+\Pi)}{\tilde{P}_a^k(P+\Pi^+)}\right)$$
$$+ \frac{\bar{h}_s^2(P)}{P_F^k(P+L)\tilde{P}_s^k(P)P_F^k(P+L')}$$
$$\times \left(\frac{\bar{h}_\rho^2(P)}{\tilde{P}_\rho^k(P)} + \frac{\bar{h}_\rho^2(P)}{\tilde{P}_\rho^k(P+0^+)} + \frac{\bar{h}_\rho^2(P+\Pi)}{\tilde{P}_\rho^k(P+\Pi)} + \frac{\bar{h}_\rho^2(P+\Pi)}{\tilde{P}_\rho^k(P+\Pi^+)}\right)$$
$$+ \frac{\bar{h}_d^2(P) f_d(P/2+L)) f_d(P/2+L')}{P_F^k(P+L)\tilde{P}_d^k(P)P_F^k(P+L')}$$
$$\left.\times \left(\frac{\bar{h}_\rho^2(P)}{\tilde{P}_\rho^k(P)} + \frac{\bar{h}_\rho^2(P)}{\tilde{P}_\rho^k(P+0^+)} + \frac{\bar{h}_\rho^2(P+\Pi)}{\tilde{P}_\rho^k(P+\Pi)} + \frac{\bar{h}_\rho^2(P+\Pi)}{\tilde{P}_\rho^k(P+\Pi^+)}\right)\right).$$

Bibliography

[1] J. Hubbard, Proc. R. Soc. London, Ser. A **276**, 238 (1963).

[2] J. Kanamori, Prog. Theor. Phys. **30**, 275 (1963).

[3] M. C. Gutzwiller, Phys. Rev. Lett. **10**, 159 (1963).

[4] K. Miyake, S. Schmitt-Rink, and C. M. Varma, Phys. Rev. B **34**, 6554 (1986).

[5] D. J. Scalapino, E. Loh, and J. E. Hirsch, Phys. Rev. B **34**, 8190 (1986).

[6] N. E. Bickers, D. J. Scalapino, and R. T. Scalettar, Int. J. Mod. Phys. B **1**, 687 (1987).

[7] P. A. Lee and N. Read, Phys. Rev. Lett. **58**, 2691 (1987).

[8] A. J. Millis, H. Monien, and D. Pines, Phys. Rev. B **42**, 167 (1990).

[9] P. Monthoux, A. V. Balatsky, and D. Pines, Phys. Rev. Lett. **67**, 3448 (1991).

[10] D. J. Scalapino, Phys. Rep. **250**, 329 (1995).

[11] N. E. Bickers, D. J. Scalapino, and S. R. White, Phys. Rev. Lett. **62**, 961 (1989).

[12] N. Bulut, D. J. Scalapino, and S. R. White, Phys. Rev. B **47**, 6157 (1993); **47**, 14599 (1993).

[13] T. A. Maier, M. Jarrell, T. Pruschke, and J. Keller, Phys. Rev. Lett. **85**, 1524 (2000).

[14] T. A. Maier, M. Jarrell, T. C. Schulthess, P. R. C. Kent, and J. B. White, Phys. Rev. Lett. **95**, 237001 (2005).

[15] D. Sénéchal, P.-L. Lavertu, M.-A. Marois, and A.-M. S. Tremblay, Phys. Rev. Lett. **94**, 156404 (2005).

[16] T. A. Maier, M. S. Jarrell, and D. J. Scalapino, Phys. Rev. Lett. **96**, 047005 (2006).

[17] T. A. Maier, A. Macridin, M. Jarrell, and D. J. Scalapino, Phys. Rev. B **76**, 144516 (2007).

[18] D. J. Scalapino, Chapter13 in "Handbook of High Temperature Superconductivity", J. R. Schrieffer and R. S. Brooks, editors, Springer (2007).

[19] H. J. Schulz, Europhys. Lett. **4** (5), 609 (1987).

[20] I. Dzyaloshinskii, Sov. Phys. JETP **66**, 848 (1987).

[21] P. Lederer, G. Montambaux, and D. Poilblanc, J. Phys. (Paris) **48**, 1613 (1987).

[22] D. Zanchi and H. J. Schulz, Z. Phys. B **103**, 339 (1997).

[23] D. Zanchi and H. J. Schulz, Europhys. Lett. **44**, 235 (1998).

[24] C. J. Halboth and W. Metzner, Phys. Rev. Lett. **85**, 5162 (2000).

[25] C. J. Halboth and W. Metzner, Phys. Rev. B **61**, 7364 (2000).

[26] M. Salmhofer and C. Honerkamp, Prog. Theor. Phys. **105**, 1 (2001).

[27] C. Honerkamp, M. Salmhofer, N. Furukawa, and T. M. Rice, Phys. Rev. B **63**, 035109 (2001).

[28] C. Honerkamp and M. Salmhofer, Phys. Rev. B **64**, 184516 (2001).

[29] A. A. Katanin and A. P. Kampf, Phys. Rev. B **72**, 205128 (2005).

[30] T. Baier, E. Bick, and C. Wetterich, Phys. Rev. B **62**, 15471 (2000).

[31] T. Baier, E. Bick, and C. Wetterich, Phys. Rev. B **70**, 125111 (2004).

[32] T. Baier, E. Bick, and C. Wetterich, Phys. Lett. B **605**, 144 (2005).

[33] H. C. Krahl and C. Wetterich, Phys. Lett. A **367**, 263 (2007).

[34] H. C. Krahl, J. A. Müller, and C. Wetterich, Phys. Rev. B **79**, 094526 (2009).

[35] H. C. Krahl, S. Friederich, and C. Wetterich, Phys. Rev. B **80**, 014436 (2009).

[36] S. Friederich, H. C. Krahl and C. Wetterich, Phys. Rev. B **81**, 235108 (2010).

[37] J. Hubbard, Phys. Rev. Lett. **3**, 77 (1959).

[38] R. L. Stratonovich, Soviet. Phys. Doklady **2**, 416 (1958).

[39] F. J. Wegner and A. Houghton, Phys. Rev. A, **8**, 401 (1973).

[40] K. G. Wilson and J. Kogut, Phys. Rep. C, **12**, 75, (1974).

[41] K. G. Wilson, Rev. Mod. Phys. **47**, 773 (1975).

[42] J. Polchinski, Nucl. Phys. B **231**, 269 (1984).

[43] C. Wetterich, Phys. Lett. B **301**, 90 (1993).

[44] C. Husemann and M. Salmhofer, Phys. Rev. B **79**, 195125 (2009).

[45] M. Salmhofer, C. Honerkamp, W. Metzner, and O. Lauscher, Prog. Theor. Phys. **112**, 943 (2004).

[46] M. Ossadnik and C. Honerkamp, arXiv:0911.5047.

[47] P. Strack, R. Gersch, W. Metzner, Phys. Rev. B **78**, 014522 (2008).

[48] C. C. Tsuei and J. R. Kirtley, Rev. Mod. Phys., **72**, 969 (2000).

[49] D. Rohe and W. Metzner, Phys. Rev. B **71**, 115116 (2005).

[50] A. A. Katanin and A. P. Kampf, Physica B **359**, 557 (2005).

[51] A. P. Kampf and A. A. Katanin, Journal of Physics and Chemistry of Solids **67**, 114 (2006).

[52] A. Katanin, Phys. Rev. B **81**, 165118 (2010).

[53] P. W. Anderson, Science **235**, 1196 (1987).

[54] E. H. Lieb and F. Y. Wu, Phys. Rev. Lett. **20**, 1445 (1968).

[55] J. Jaeckel and C. Wetterich, Phys. Rev. D **68**, 025020 (2003).

[56] R. J. Birgeneau *et al.*, Phys. Rev. B **39**, 2868 (1989).

[57] S.-W. Cheong *et al.*, Phys. Rev. Lett. **67**, 1791 (1991).

[58] B. J. Sternlieb, J. M. Tranquada, G. Shirane, M. Sato and S. Shamoto, Phys. Rev. B **50**, 12915 (1994).

[59] F. Yuan, S. Feng, Z. B. Su and L. Yu, Phys. Rev. B **64**, 224505 (2001).

[60] J. Lorenzana and G. Seibold, Phys. Rev. Lett. **89**, 136401 (2002).

[61] M. Fujita, H. Goka, K. Yamada, J. M. Tranquada and L. P. Regnault, Phys. Rev. B **70**, 104517 (2004).

[62] G. Seibold and J. Lorenzana, Phys. Rev. Lett. **94**, 107006 (2005).

[63] W. Metzner, J. Reiss, and D. Rohe, Phys. Status Solidi B **243**, 46 (2006).

[64] J. Reiss, D. Rohe, and W. Metzner, Phys. Rev. B **75**, 075110 (2007).

[65] M. Salmhofer, *Renormalization: An Introduction*, Springer, Heidelberg (1999).

[66] J. Berges, N. Tetradis, and C. Wetterich, Phys. Rep. **363**, 223 (2002).

[67] J. M. Pawlowski, Annals Phys. **322**, 2831 (2007).

[68] H. Gies and C. Wetterich, Phys. Rev. D **65**, 065001 (2002).

[69] S. Floerchinger and C. Wetterich, Phys. Lett. B **680**, 371 (2009).

[70] K.-U. Giering and M. Salmhofer, in preparation.

[71] D. F. Litim, Phys. Lett. B **486**, 92 (2000).

[72] D. F. Litim, Phys. Rev. D **64**, 105007 (2001).

[73] H. C. Krahl, *PhD thesis*, Heidelberg, (2007).

[74] S. Chakravarty, B. Halperin, and D. Nelson, Phys. Rev. B **39**, 2344 (1989).

[75] C. Wetterich, Z. Phys. C **57**, 451 (1993).

[76] J. M. Kosterlitz and D. J. Thouless, J. Phys. C. **6**, 1181 (1973).

[77] M. Grater and C. Wetterich, Phys. Rev. Lett. **75**, 378 (1995).

[78] G. v. Gersdorff and C. Wetterich, Phys. Rev. B **64**, 054513 (2000).

[79] A. Eberlein and W. Metzner, Prog. Theor. Phys. **124**, 471 (2010).

[80] P. Jacubczyk, W. Metzner, and H. Yamase, Phys. Rev. Lett. **103**, 220602 (2009).

I want morebooks!

Buy your books fast and straightforward online - at one of world's fastest growing online book stores! Environmentally sound due to Print-on-Demand technologies.

Buy your books online at
www.morebooks.shop

Kaufen Sie Ihre Bücher schnell und unkompliziert online – auf einer der am schnellsten wachsenden Buchhandelsplattformen weltweit! Dank Print-On-Demand umwelt- und ressourcenschonend produziert.

Bücher schneller online kaufen
www.morebooks.shop

KS OmniScriptum Publishing
Brivibas gatve 197
LV-1039 Riga, Latvia
Telefax +371 686 204 55

info@omniscriptum.com
www.omniscriptum.com

Printed by Books on Demand GmbH, Norderstedt / Germany